THEORY AND PRACTICE IN ARISTOTLE'S NATURAL SCIENCE

Aristotle argued that in theory one could acquire knowledge of the natural world. But he did not stop there; he put his theories into practice. This volume shows how Aristotle's natural science and philosophical theories shed light on one another. The contributors engage with both biological and non-biological scientific works and with a wide variety of theoretical works, including *Physics*, *Generation and Corruption*, *On the Soul*, and *Posterior Analytics*. The chapters focus on a number of themes, including the sort of explanation provided by matter; the relationship between matter, teleology, and necessity; cosmic teleology; how an organism's soul and faculties relate to its end; how to define things such as sleep, void, and soul; and the proper way to make scientific judgments. The resulting volume offers a rich and integrated view of Aristotle's science and shows how it fits with his larger philosophical theories.

DAVID EBREY is Assistant Professor of Philosophy at Northwestern University, Illinois.

THEORY AND PRACTICE IN ARISTOTLE'S NATURAL SCIENCE

EDITED BY

DAVID EBREY

Northwestern University, Illinois

CAMBRIDGE
UNIVERSITY PRESS

CAMBRIDGE
UNIVERSITY PRESS

University Printing House, Cambridge CB2 8BS, United Kingdom

Cambridge University Press is part of the University of Cambridge.

It furthers the University's mission by disseminating knowledge in the pursuit of education, learning and research at the highest international levels of excellence.

www.cambridge.org
Information on this title: www.cambridge.org/9781107055131

First published 2015

A catalogue record for this publication is available from the British Library

ISBN 978-1-107-05513-1 Hardback

Contents

PART III METHODOLOGY

Notes on contributors

DAVID EBREY is Assistant Professor at Northwestern University. He has written articles on Plato's *Meno* and *Phaedo* and on Aristotle's natural science and logic.

ALAN CODE is Ward W. and Priscilla B. Woods Professor of Philosophy at Stanford University. He is the author of articles on Aristotle's metaphysics, science, and logic.

JESSICA GELBER is Assistant Professor at the University of Pittsburgh. She has written articles on metaphysical issues in Aristotle's science.

MARGARET SCHARLE is Associate Professor at Reed College. She has written articles on Aristotelian natural philosophy.

DEVIN HENRY is Associate Professor at the University of Western Ontario. He is the author of several articles on topics in Aristotle's philosophy of science (including classification, teleology, inheritance, and sexism) as well as Plato's late epistemology. He is the editor, along with Karen M. Nielson, of *Bridging the Gap Between Aristotle's Science and Ethics* (2015).

THOMAS K. JOHANSEN is Professor of Ancient Philosophy at Oxford University and Tutorial Fellow at Brasenose College. He is the author of *Aristotle on the Sense-Organs* (1998), *Plato's Natural Philosophy: A Study of the Timaeus-Critias* (2004), and *The Powers of Aristotle's Soul* (2012).

CHRISTOPHER FREY is Assistant Professor at the University of South Carolina. He is the author of articles on Aristotle's natural philosophy and metaphysics and on contemporary philosophy of perception and mind. He is writing a book on the role of the soul as a principle of life in Aristotle.

MONTE RANSOME JOHNSON is Associate Professor at the University of California, San Diego. He is the author of *Aristotle on Teleology* (2005) and of articles on Democritus, Aristotle, and Lucretius. He is currently working with D. S. Hutchinson on a reconstruction of Aristotle's lost work the *Protrepticus.*

DAVID SEDLEY is Emeritus Laurence Professor of Ancient Philosophy at the University of Cambridge and a Fellow of Christ's College, Cambridge. He is the author of *The Hellenistic Philosophers* (1987, with A. A. Long), *Plato's* Cratylus (2003), *The Midwife of Platonism: Text and Subtext in Plato's Theaetetus* (2004), and *Creationism and its Critics in Antiquity* (2007), as well as the editor of four collections of essays.

SEAN KELSEY is Associate Professor at the University of Notre Dame. He has written articles on Aristotle's *Physics* and Plato's *Phaedo* and *Republic.*

ANDREA FALCON is Associate Professor at Concordia University. He is the author of *Corpi e Movimenti: Il De Caelo di Aristotele e la sua tradizione nel mondo antico* (2001), *Aristotle and the Science of Nature: Unity without Uniformity* (2005), and *Aristotelianism in the First Century* BCE: *Xenarchus of Seleucia* (2011).

MARISKA LEUNISSEN is Associate Professor at the University of North Carolina at Chapel Hill. She is the author of *Explanation and Teleology in Aristotle's Science of Nature* (2010) and the editor of *Aristotle's Physics, A Critical Guide* (forthcoming).

CHAPTER I

Introduction

David Ebrey

Aristotle thought that it was possible to acquire scientific knowledge of the natural world but that none of his predecessors had the philosophical theories needed to develop such knowledge. One task he took for himself, then, was to provide a foundation for natural science. However, he did not simply argue that in principle one could acquire this sort of scientific knowledge; he put his theories into practice. More than 750 pages of his scientific works survive. These works are fascinating in their own right and have the power to help us understand central features of his natural philosophy. His practice sheds light on his theory and this theory, in turn, sheds light on his practice. His scientific works can, among other things, help us better understand his accounts of matter, necessity, teleology, definition, and proper scientific methodology.

Aristotle wrote a number of works that, in different ways, provide a theoretical foundation for the study of the natural world. His *Physics* is devoted precisely to this task. The first book of *Generation and Corruption* provides general accounts of different types of change, which Aristotle thought were applicable across different parts of the natural world. By contrast, his *Posterior Analytics* has a broader scope than either of these works: it is meant to explain methodology relevant to any science, not necessarily natural science. Thus, for example, the *Posterior Analytics* is supposed to apply equally to mathematics, which Aristotle does not consider a natural science. The first book of the *Parts of Animals* provides a foundation for natural science, but with an emphasis on the study of biology. Similarly, the *de Anima* (*On the Soul*) is an inquiry into and account of the soul and its activities, which is directly relevant to the biological sciences, since Aristotle thinks that the soul is the principle of life.

Aristotle's scientific works are an ambitious attempt to explain most aspects of the natural world. The non-biological works take us from the whole cosmos (*de Caelo* I and II) to the elements (*Generation and Corruption* II, *de Caelo* III and IV, *Meteorology* IV), and through meteorology

(*Meteorology* I–III). The lion's share of the works is biological, in particular animal biology. Aristotle's *History of Animals* is the longest work in his corpus; it is now widely thought to provide a systematic account of animal parts and features that could be drawn on in causal treatises. By contrast, *Parts of Animals*, *Generation of Animals*, *Movement of Animals*, and *Progression of Animals* are each causal treatises: they each set out to determine the explanation for why animal parts have the features that they do, why animals generate in the way that they do, etc. A third and final group of works are known as the *Parva Naturalia*, which Aristotle says comes after his examination of the soul (436a1–5); these discuss topics such as memory and sleep. These are causal treatises and, at the same time, more closely related to *de Anima* than the other causal biological works. They are a reminder that Aristotle himself does not operate with a strict division between theory and practice and so we should be careful not to treat it as a fundamental divide in his works.

Most people come to Aristotle from an interest in the history of philosophy, rather than the history of science, and so Aristotle's scientific works have received less attention than his other works. However, over the last thirty years scholars have steadily increased their interest in his scientific works, in part because these can help us answer long-standing questions about his philosophical theories, making this a very exciting time in Aristotle scholarship. Despite this recent interest, the last volume of essays whose scope included all of Aristotle's natural science is more than thirty-five years old: *Articles on Aristotle, Vol.* I*: Science*, eds. Barnes, Schofield, and Sorabji. Since then groundbreaking work has been done, especially on Aristotle's biology and its relation to the scientific method in the *Posterior Analytics*. The most important volume on this is *Philosophical Issues in Aristotle's Biology*, edited by Gotthelf and Lennox, which includes seminal essays at the foundation of much of the current research on Aristotle's biology.

This volume builds on the last thirty years of scholarship, broadening it in two ways: (i) it includes chapters on both Aristotle's biological and nonbiological works and (ii) it includes a number of chapters that emphasize connections between scientific treatises and foundational treatises other than the *Posterior Analytics*, such as the *Physics*. The volume does not aim to provide a comprehensive approach to the topic, but rather to be a collection of important new work. Each chapter raises a new and important issue or proposes a new way to approach a classic question. One surprising feature of the contributions is that, despite the diversity of questions and approaches, they intersect to an unusually high degree. This overlap was not by design; these simply are the issues that scholars are excited about and find most pressing. The chapters naturally fall into three groups, organized around three key topics: matter, teleology, and methodology. Some address more than one topic, so those interested in one of these topics should read the descriptions below to see if chapters in other sections are relevant. There are a number of other recurring topics that could have been used to organize the volume, such as necessity, definition, form, soul, and cosmology.

Aristotle was the first person to develop the concept of matter; he thought matter was vital for a proper understanding of the natural world. The first three chapters in this volume tackle fundamental questions about matter. What sort of explanation does matter provide? What sorts of things does Aristotle identify as matter? Does he treat matter as fundamental in his explanations? How does matter relate to so-called material necessity? And what is the role of matter in scientific definitions?

In "The 'Matter' of Sleep," Alan Code uses Aristotle's *Metaphysics* H4, to shed light on what the material cause of sleep is in the *de Somno* (*on Sleep*), a question that has received a number of very different answers. According to H4 non-substances do not have matter but have an analogue to matter, the central features of which, Code argues, are that it underlies a given change and it is receptive to the relevant opposites in this change. According to Code current interpretations of *de Somno* have not identified the material cause as the primary thing that undergoes the given change, sleep. In H4 Aristotle says that the matter-analogue is not the whole animal, but the primary part of the animal that undergoes sleep. Code argues that in the *de Somno* the heart is the relevant part for blooded animals, and hence it is the matter-analogue. He uses this account to show how Aristotle in the H4 passage extends his four causes from sensible substances to the attributes of substances, such as sleep.

In "Are Facts about Matter Primitive?" Jessica Gelber argues against recent scholars who claim that Aristotle in his biology treats "facts about matter" – facts such as the degree of heat or amount of fluidity in an organism's material constitution – as explanatorily basic or primitive. Gelber presents three considerations for rejecting this recent consensus, which she claims is in tension with Aristotle's general commitment to the causal and explanatory priority of form over matter. First, she argues that Aristotle in fact does explain certain facts that others have described as unexplained, such as an organism's degree of heat, dryness, and fluidity. Second, she argues that in certain cases, such as human intelligence, it would be quite implausible to suppose that Aristotle considers an explanation in terms of degrees of heat to be a primitive starting point. Finally, Gelber argues that in order for facts about matter to be the primitive basis for explanations

Aristotle would need a specific way of thinking about relevant causal processes. But Aristotle does not characterize the relevant causal processes in the requisite way in the *Generation of Animals*.

In "Blood, Matter, and Necessity," I argue that in the *Parts of Animals* Aristotle thinks of matter and necessity in a way that makes it highly misleading to describe his views in terms of "material necessity," as most recent scholars do. Aristotle rarely discusses matter in the *Parts of Animals* and when he does he typically identifies blood as matter. This, I argue, is because he thinks of matter as what things properly come to be from and he thinks that the body concocts blood precisely to be what things come to be from. Blood, far from being linked to necessity, is connected to divergent possibilities: it can come to be, nourish, or grow different parts of the body. By contrast with his occasional references to matter, Aristotle frequently explains things in terms of necessity in the *Parts of Animals*. I argue that he does not distinguish between different kinds of necessity in his explanations and that in some cases the necessity has nothing to do with matter. Aristotle groups all cases of necessity together; what he thinks is important is how these explanations differ from teleological explanations.

Aristotle famously thinks that we should understand natural entities and their changes in terms of each thing's end or goal (*telos*). The next four chapters in the volume address crucial questions about Aristotle's teleology: How does he integrate his teleological explanations with those in terms of necessity? How should we understand Aristotle's distinction between two different kinds of end, that for the sake of which and the beneficiary? Does teleology only work at the level of individual organisms, or does it also provide explanations at the level of kinds? How do an organism's soul and faculties relate to its end?

In "'And These Things Follow': Teleology, Necessity, and Explanation in Aristotle's *Meteorologica*," Margaret Scharle argues that while teleology is not mentioned in the *Meteorologica*, it is implicitly presupposed there. In particular, the *Meteorologica*'s explanations depend on Aristotle's view that there are two elemental cycles, an earth–fire cycle and an air–water cycle, which Aristotle ultimately explains in terms of teleological processes outside of the *Meteorologica*. Scharle argues that Aristotle's procedure in explaining meteorology is just what one would expect from *Parts of Animals* I.1 and his practice in the biological works: he is examining the necessary by-products of teleological processes. The difference between biological and meteorological cases, she argues, is that in biology the organism can make use of necessary by-products to further the organism's ends. By contrast, there is no way for the elements to make use of necessary by-products to further their ends because they are entirely passive and have no parts.

In "Aristotle on the Cosmological Significance of Biological Generation," Devin Henry argues that in *Generation of Animals* II.1 Aristotle does not limit teleology to what will benefit a specific organism, as is commonly thought. Instead, Aristotle thinks that there must be reproduction for the benefit of the continuous generation of the whole species, where this benefit to the species cannot be reduced to the benefit to individual organisms. Henry argues for this, in part, by drawing on parallels between Aristotle's argument in *GA* II.1 and his argument in *Generation and Corruption* II.10 for the necessity of cyclical generation. His chapter ends by suggesting a way that this interpretation can help us make headway on a very difficult argument in *GC* II.11.

In "The Two Kinds of End in Aristotle: The View from the *de Anima*," Thomas Johansen argues that Aristotle's distinction between two kinds of end, frequently dismissed as an aside, in fact helps us understand how different faculties are related to one another for the overall benefit of an organism. One kind of end is that for the sake of which a change or activity takes place; the other kind is the beneficiary of the change or activity. Johansen provides a general account of this distinction, including its Platonic precedent, and then considers its role in the *de Anima*. He argues that it explains how the soul's faculties can be defined strictly independently of one another (in terms of end as that for the sake of which), while at the same time these faculties are closely related to one another, in a way that is central to an animal's nature (in terms of end as beneficiary). The faculties have strict goals, but they also benefit the other faculties and the organism as a whole.

In "Two Conceptions of Soul in Aristotle," Chris Frey argues against the dominant interpretation of soul in Aristotle, according to which it is understood in terms of a group of capacities (which is the view defended by Johansen in this volume), and in favor of a more unified conception of the soul, according to which it is the single end of all of the organism's activities. Frey argues for this reading by examining how Aristotle accomplishes the task he sets for himself in *de Anima* II.2: to come up with a more scientific definition of the soul. On his account, Aristotle unifies the soul by creating a hierarchy of souls, with the lower souls, for example, the nutritive soul, present potentially (*dunamei*) in the higher soul. Frey uses Aristotle's account of mixture to shed light on how to understand this.

Over the last thirty years, scholars have made significant advances by using Aristotle's methodology in the *Posterior Analytics* to shed light on his natural science. The last four chapters help make this relationship more reciprocal by using Aristotle's natural science to illuminate central topics in his *Posterior Analytics*: How are subordinate and superordinate sciences related to one another? And what are the right sorts of definitions to use in natural science? The contributions also examine important methodological issues that have been overlooked or misunderstood, in part because they are not discussed in the *Posterior Analytics*: What are the basic requirements for exercising scientific judgment? And how should the scientist proceed when lacking sufficient evidence?

In "Aristotle's Architectonic Sciences," Monte Johnson addresses two puzzles about the autonomy and subordination of sciences. He does so, in part, by comparison and contrast with Plato's views in the *Statesman* and by bringing to bear underappreciated evidence from Aristotle's *Protrepticus*. The first puzzle is that in the *Nicomachean Ethics* Aristotle treats political science as the most architectonic science whereas in the *Metaphysics* he treats metaphysics (i.e., first philosophy) as most architectonic. Johnson argues that Aristotle's considered view is that metaphysics is the most architectonic, and, in general, that Aristotle views productive sciences as subordinate to practical sciences and these, in turn, as subordinate to theoretical sciences. Johnson's second puzzle is how Aristotle can view all sciences as subordinate to a master science, or even any science as subordinate to any other, given his views on the independence of the sciences. He creates a model for how to do so, giving examples from Aristotle's practice within meteorology and politics, and showing how in both cases these sciences draw on theoretical claims from metaphysics, mathematics, and natural science.

In "Varieties of Definition," David Sedley uses Aristotle's multiple definitions of void in *Physics* IV.7 to shed new light on an important and controversial debate about the *Posterior Analytics*: Does Aristotle think that there is a separate class of nominal definitions, that is, definitions that do not presuppose the existence of the thing defined? Many scholars claim to find evidence in *Posterior Analytics* II.10 for such definitions. If nominal definitions were a distinct group in II.10, they would be distinct from definitions that correspond to the conclusion of a syllogism. However, Sedley notes that if ever there were a place for Aristotle to provide a nominal definition, it would be for void, given that he thinks that there is no such thing as void. Instead, Sedley argues that Aristotle provides multiple definitions for void using syllogisms. Moreover, these definitions, if anything, point to a way in which even void can be said to exist.

In "Empty Words," Sean Kelsey examines Aristotle's surprising claims that his predecessors are using mere empty words, with no conviction about what they are saying. Kelsey uses Aristotle's ethical works to shed light on these and similar accusations that are found across his scientific works. He argues that Aristotle does not think that his predecessors have merely failed to hit upon the truth, but rather that they are engaging in a sort of imitation and pretend version of the genuine search for truth. To make matters worse, Aristotle thinks they do not even realize that they are so far from the genuine search for truth. Kelsey argues that Aristotle thinks that this is a depressingly common mistake, not a rare error. In particular, Aristotle thinks it is easy to let one's focus on reasoning and argumentation lead to focusing on one's dialectical opponents, rather than attending to the empirical facts appropriate to the subject matter at hand and using these to discover the truth.

In "The Scientific Role of *Eulogos* in Aristotle's *Cael* II 12," Andrea Falcon and Mariska Leunissen argue that Aristotle is engaged in natural scientific inquiry when, in his natural scientific works, he uses claims that he describes as "reasonable" (*eulogos*) or unreasonable. They argue he is not merely engaged in dialectic, as is sometimes suggested. Aristotle uses these claims about what is reasonable as part of an overall method for providing the best possible scientific explanations in cases where we lack sufficient empirical evidence needed for certainty. Falcon and Leunissen apply this account of Aristotle's use of reasonable claims to *de Caelo* II 12, arguing that, despite the difficulties of the puzzles presented in II 12 and despite Aristotle's modest goals in responding to these puzzles, his responses are proper scientific explanations, not merely dialectical ones.

Together, these chapters challenge the orthodoxy on Aristotle's views on matter, clarify non-standard types and uses of teleology, and shed light on underappreciated parts of his scientific methodology. Moreover, they bring to the fore his cosmology, put forward new ways to understand the unity of the soul, and help us understand how necessity applies to the elements. They present groundbreaking work on Aristotle's natural philosophy and provide a model for how to integrate our understanding of Aristotle's theory and practice.

PART I

Matter

CHAPTER 2

The 'matter' of sleep

Alan Code

Matter, metaphysics and the causes of natural things

Metaphysics H4 starts with a discussion of material substance (οὐσία ὑλική), and the matter that is appropriate for each kind of thing (οἰκεία ἑκάστου), and ends with a discussion of things that do not even have matter, but have something else instead. Sleep belongs to the latter group, and is brought in at the end of a section of text in which Aristotle talks about the correct way to inquire into the causes of natural things. Before looking specifically into the matter, or matter-analogue, for sleep, we should first consider the context within which that topic comes up.

For Aristotle, sleep is something that exists by nature, but is not a substance. When he discusses the right way to state causes for items of this sort in *Metaphysics* H4 he says that they do not have matter. They do, however, have a cause that plays an analogous role, and he goes on to say what this is and how it is connected with two other types of cause. There are a number of reasons that this is important. One is that there are scholarly controversies about Aristotle's views on the material cause of sleep in the *de Somno*, and anger in *de Anima* I.I, and more generally about the material cause for the attributes he classifies as common to body and soul.[1] I will touch on this briefly below, since the passage in *Metaphysics* H4 that expresses a view about the causes of such things, including the cause that corresponds to the matter of a generable substance, puts constraints on possible candidates for that role. However, it is by no means obvious how a hylomorphic natural science should analyze or explain natural entities that are not themselves composite substances, but rather are the attributes of these substances. Fortunately, this passage in the *Metaphysics* provides a principled perspective as to how this is to be done.

[1] For Aristotle's use of this phrase in connection with sleep see Morel 2006.

The general methodological remarks about the proper statement of causes are introduced immediately after he has given some examples of various kinds of matter, and examples that relate something's matter to an efficient cause. These examples illustrate ways in which the appropriate matter for X is located within a larger network of causes. First, he uses some examples to show how the appropriate matter for X is the proximate matter for X (what he will go on to describe as a *'nearest'* cause for the thing.)[2] He does this by giving examples of cases in which there is different proximate matter for different things, as well as by giving examples that illustrate a hierarchical conception of matter that involves both upwards composition and downwards decomposition, followed by upwards composition. Next he brings in a different type of cause, the efficient cause, together with some new examples that illustrate additional constraints on the matter for X.[3] This move brings in a further dimension to his discussion of the way in which something is the appropriate matter for X. The matter's suitability for its role is not independent of the efficient cause that brings it about. Considerations concerning what that efficient cause is, or is not, able to produce from various kinds of matter make it evident, among other things, that nothing could be the appropriate matter for X unless an efficient cause could make an X from it.

Although these two factors have different causal roles, these are roles that they do not play independently of each other. The point is not simply that the matter needs something else, an efficient cause, to produce X; nor is the point simply that the efficient cause requires something else, the matter, if it is going to bring about an X. The language of *agency* is used here to talk about the efficient cause being able, or not able, to make something out of something.[4] For instance, the efficient cause of a saw *produces*, or is *not able to produce*, a saw *from* certain types of matter. A saw could not be produced from matter such as wood or wool. The appropriate matter for a saw is something that is capable of being acted upon by the efficient cause in such a way that a saw is produced out of that matter. These causes are related as agent and patient, and for something to be the matter of X it must be capable of being affected in such a way that an X is made from it, and the efficient cause needs to be capable of acting on that matter in such a way as to bring about that result.[5]

[2] *Metaph.* H4, 1044a18–25. [3] *Metaph.* H4, 1044a25–32. [4] *Metaph.* H4, 1044a29; a30.

[5] At the end of the chapter, in discussing the causes of sleep, he uses πάσχειν in connection with the item affected by the efficient cause; see *Metaph.* H4, 1044b11, 16, 19.

This is true not of the *remote* efficient or material causes, but rather is a requirement for the '*nearest*,' or proximate causes. The proximate efficient cause cannot be understood in its role *as* the efficient cause to X independently of the proximate matter being understood in its role *as* the material cause of X. The former is a cause by being the *agent* that acts on the matter in such a way as to produce X from it; and the latter is a cause by being the *patient* that is affected by the efficient cause in such a way that X is produced from it. In general, one way of being a cause is to be what acts in that way; the other way of being a cause is to be what is affected in that way. Each of these two items is a proximate cause for X, and each is a cause in a different way, but neither makes its causal contribution independently of the other.

It is at this point, after discussing material substance in connection with the efficient cause, that Aristotle generalizes in a way that brings in whatever other types of causes something might have:

> In fact, whenever one inquires into the cause, it is necessary to state all of the possible causes, since causes are spoken of in many ways.[6]

Not only are the material and efficient causes mutually interdependent, but since there are many (i.e., four) ways in which something can be said to be a cause, it is necessary to state *all* of the possible causes. Aristotle proceeds to illustrate this with the example of the causes of a human. In doing so he also goes on to make it clear that the procedure he describes is not simply one for *saying* what the causes of something are, but is also being put forward as the right way to *pursue* causes:

> concerning the natural and generable substances, one must go after them in this way if one is to go after them rightly, since the causes are indeed these and this many, and it is necessary to know the causes.[7]

For Aristotle's example of a generable substance, a human, he poses a series of four questions corresponding to each of his four causes: What is the cause as matter (ὡς ὕλη)? What is the cause as what effects change (ὡς κινοῦν)? What is the cause as the form (ὡς τὸ εἶδος)? And what is the cause as that for the sake of which (ὡς οὗ ἕνεκα)? The first question is itself answered with the question 'is it the menstrual fluid [τὰ καταμήνια]?' The

[6] *Metaph.* H4, 1044a32–34. [7] *Metaph.* H4, 1044b3–5.

second is answered with the question 'is it the seed [τὸ σπέρμα]?'[8] The third and fourth questions are answered with the replies 'essence [τὸ τί ἦν εἶναι],' and 'end [τὸ τέλος],' respectively, followed by the suggestion that perhaps both are the same.

As in the discussion of material substance in the earlier parts of *Metaphysics* H4, here too Aristotle is identifying matter with *that from which something comes to be*. It falls outside of the scope of the present chapter to attempt a full discussion of how this is connected with the general definition(s) of soul in *de Anima* II.1, and how it bears on issues concerning the soul as the form of the body. These passages in *Metaphysics* H4 treat the matter *as that from which something comes to be*, and do not bring in some other conception of the matter of a generable substance according to which the matter exists at all and only those times at which it possesses the form, or soul, of the animal. Furthermore, like his treatment in *Metaphysics* H4, the way Aristotle discusses the matter of an animal in the following chapter continues to treat it as that from which something comes to be. There the matter of an animal is the matter of the corpse that comes into being when the animal perishes.[9] The purpose of this chapter is not to resolve issues about how to relate the matter of a mature specimen of a kind to the matter for its generation, but simply to explicate the way in which *Metaphysics* H4 discusses the material cause, and to do so with an eye to what it has to say about the way in which sleep has an analogous cause.

When the H4 passage turns to the case of *eternal* natural substances, there is no discussion at all of the efficient, formal or final causes, and the only remarks made concern the material cause:

> But in the case of substances that, though natural, are eternal, another treatment [is appropriate]; for perhaps [ἴσως] some do not have matter, or not such [matter], but only [matter] that changes with respect to place.[10]

[8] It is worth pointing out that he does not go on to affirm that the matter is the menstrual fluid. When looking into the question 'what is the matter?', an answer such as 'fire' or 'earth' fails to cite what is 'nearest,' and instead one should look for what is distinctive (ἴδιον) (*Metaph.* H4, 1044a33–b3). The menstrual fluid, a concoction of the female's blood, is distinctive to each type of animal. However, this does not show that menstrual fluid is compatible with the view that what becomes the animal is the matter for both the menstrual fluid (which perishes) and the animal (which comes to be). In the *Generation of Animals* the matter of a blooded animal is said to be in the composition (σύστασις) of the menses (*GA* I.19, 727b31–33); alternatively, the nature of the menstrual fluid (*GA* I.20, 729a32–33). It is what contains all of the parts of the animal in capacity (δυνάμει), though not in actuality (ἐνεργείᾳ) (*GA* II.3, 737a23–24).

[9] *Metaph.* H5, 1045a1–2.

[10] *Metaph.* H4, 1044b6–8.

The account that one gives of such substances should also cite all of the possible causes, and the most proximate ones at that, but all that we are told here is how the causal account one gives for them will differ from the account of a generable substance in one respect – in whether, and if so how, to cite matter. Perhaps there are *some* natural substances that do not even have matter. This may be leaving open, at least for now, the question as to whether there are *natural* substances that are not only ungenerable, but also not even subject to change at all. For instance, although Aristotle himself would not accept such a view, perhaps there are certain mathematical objects, such as the shapes of heavenly bodies, that exist by nature and also have the ontological status of substances.[11] However, for no eternal substance could there be the matter *from which it comes to be*, and so eternal, natural substances do not have the kind of matter that generable substances have. Hence they could not be composites of matter from which they were generated and a form acquired as the completion of their generation. However, heavenly bodies do engage in eternal circular motion, and hence there is at least some way in which such things have matter. They have matter for change of place,[12] but for all that they are not hylomorphic composites in the way that is distinctive of generated substances.

Additionally, Aristotle does not analyze eternal, celestial substances into different functional parts, or parts composed of different materials or mixtures of materials. Regardless of whether this is due to their nature (as he thinks it is), or to limitations in our knowledge of them, or even to both, for heavenly bodies we do not inquire into such causes. This in turn has consequences for the way in which we can seek causes for their motion and other attributes. In particular, although an eternal substance is itself a subject for attributes, we are not able to investigate these attributes in the same way that we ought to investigate the attributes of perishable substances. For perishable substances, often we are able to proceed by identifying the part in virtue of which it has the attribute, as well as a condition that part is in when the substance has the attribute, and additionally an efficient cause for that part's being in that condition. However, this is not the procedure he specifies for the search for the efficient cause of an explicable attribute of an eternal natural substance. Even though, for instance, the matter of a heavenly body is such as to make that body a subject of circular motion, he does not propose that one explains that circular motion by

[11] Or perhaps this leaves open (for now) the possibility that the source of change for the outermost heavenly sphere is a natural, though incorporeal and unchangeable, substance.

[12] For which see *Metaph.* H1, 1042b6.

identifying some part or parts out of which the subject is constituted, and some attributes or motions of such parts.

After his remarks about matter, or the lack of it, for eternal natural substances, he then turns to consider the causes of natural non-substances. Here he does assign to both eclipses and sleep a cause that somehow corresponds to the cause 'as matter' [ὡς ὕλη]. However, there can be no doubt that he also says that it is *not* matter:

> Nor indeed is there matter for those things that exist by nature but are not substances, instead 'that which underlies' [τὸ ὑποκείμενον] is substance.[13]

Let us call the cause of sleep that is 'not matter,' but is what sleep has instead, its 'matter-analogue.' It is in the final section of *Metaphysics* H4 that he discusses the causes of sleep in particular.[14]

In the next two sections I consider two questions about this cause, the 'matter-analogue.' The first has to do with *why* he says that natural attributes like sleep lack matter; the second has to do with *what a natural attribute like sleep has instead of matter*. Both questions are concerned with the way in which Aristotle's hylomorphism, as well as his doctrine of 'four causes,' handles items that are not material objects. In his ontology there are natural substances of various kinds, as well as attributes of natural substances, and both get explanations within the general hylomorphic framework of natural science.

Why does Aristotle deny that there is matter for natural phenomena such as eclipses and sleep?

We are familiar with many of the attributes that in various ways depend upon generable, natural substances. This includes such things as colors and psychological properties. Just as natural substances themselves have causes of various types, so too their natural attributes have causes. Although metaphysics is not the science that searches for the causes that natural things have insofar as they are natural, it is the discipline that explores more general questions about their causes insofar as they are beings, as well as questions about the mode of being they enjoy. An attribute such as sleep is an integral part of the cognitive and perceptual life of animals. As we are about to see, there is some controversy as to whether at the time Aristotle wrote *Metaphysics* H4 he accepted the view that sleep has

[13] *Metaph.* H4, 1044b8–9. [14] *Metaph.* H4, 1044b15–20.

a final cause, and is both necessary and beneficial for the preservation of the animal's life, where its life is both an end and a good. However, unless there is strong evidence that he had rejected such teleological explanations of sleep, which is prima facie unlikely given the centrality of teleology in his biological works, this unified structure of purposeful, goal-directed activity could itself be studied by a science of being. Nonetheless, the teleology is not brought up or discussed in *Metaphysics* H4 in connection with natural attributes.

For Aristotle natural science has its own principles, and is the appropriate locus for scientific inquiry into the sensible world. That said, it is also possible to look into the positive results of the causal investigation of nature from the more comprehensive and synoptic perspective of a metaphysical consideration of their mode of being. It is from this more general standpoint that he considers metaphysical questions about the unity of matter and form in *Metaphysics* H, and it is in this text that he also makes the claim that natural attributes do not have matter. As we have already seen, in the case of the eternal natural substances, although he denies that they have matter of the sort that generable substances possess, he nonetheless allows that they have matter in some other way – matter for change of place. Additionally, in the case of natural attributes, his denial that they have matter may not be intended to deny that there is *any* way in which they could be said to have matter.

Attributes, however, unlike celestial bodies, are not the kinds of things that change, so they could not have matter for change of place, or for any other kind of change. By way of contrast with their attributes, material objects themselves are the kinds of things that change in various ways, and with the exception of those that are eternal, they also are the kind of things that both come to be and perish. On the other hand, unlike these material composites that are generated out of some underlying matter, their forms and attributes do not come to be, but rather come to be in something else.[15]

This is, in fact, very closely related to the kind of claims he makes about opposites and forms in general in the beginning of *Metaphysics* H5, directly after his discussion of the causes of sleep:

> And since some things are and are not without coming to be and passing away, for instance points, if indeed they are, and in general the forms (for what is white does not come to be, but rather the wood [comes to be] white if everything that comes to be comes to be out of something X and [comes to be] something Y) . . .[16]

[15] *Metaph.* Z8, 1033b5–16. [16] *Metaph.* H5, 1044b21–24.

Sleep and wakefulness are also opposites, just as white and dark, and things such as these neither come into being nor pass away. Rather, they either are or are not because of a change in something else. When some wood changes in such a way that it comes to be white, the white color that it comes to possess does not also come into being. It exists when the wood comes to be white, but does so without itself coming to be, but rather because of the way the wood changes.

Speaking quite generally, everything that comes to be comes to be something Y and comes to be out of something X, regardless of whether this is the generation of a substance, or instead some non-substantial change in some substance that already exists. Even in those cases of non-substantial change, there is some item from which something comes to be:

> not all opposites [τἀναντία] would come to be out of each other, but in different ways a white man [comes to be] from a dark man and white [comes about] from dark. Nor does everything have matter, but [rather] those things of which there is coming to be and that change into one another [have matter]. But those things that are or are not without change, these do not have matter.[17]

For the change in question, the something 'from which' a white man comes to be is a dark man. He does not identify the matter of a white man with the dark man that perishes when the white man comes to be. However, the man that is first dark, and later white, is a subject for opposites, and plays a role comparable to matter in this change. Even so, the attribute whiteness is not produced out of its matter, nor is a substantial form, nor is any other item that is or is not without itself coming to be or passing away.

There is a way in which the white man and the dark man are opposites,[18] and these items do come to be from and change into each other. However, the attributes whiteness and darkness do not change into each other, and these are items that are or are not without coming to be or passing away. Likewise, sleep and wakefulness are opposites, and they too either are or are not without coming to be or passing away. Since nothing that is or is not without generation and perishing has matter, he takes it that it follows that there is no matter for sleep.

This explains the grounds for his denial that natural attributes have matter. If we were to confine our attention to *Metaphysics* H4–5 there would be no reason to think that there is another way in which such attributes could be said to have matter. Nonetheless, what he is concerned

[17] *Metaph.* H5, 1044b24–29. [18] See Ross 1924 on 1044b25.

here to deny, both in the case of eternal natural substances and in the case of natural attributes, is mainly that they have matter of the sort found for generable substances. These texts are silent on whether or not the term 'matter' could be applied in some other way to the cause for an attribute that I am calling the 'matter-analogue.'

However, according to *Metaphysics* Λ4 there is a way in which generable items (both natural substances and artifacts) have matter, and their attributes do not, and there is yet another way in which even their attributes have matter:

> In one way the causes and principles are different for different things, but in another way, should one speak comprehensively and analogically, they are the same for all.[19]

This text goes on to specify what the principles are that are the same for all things. There are three principles that are elements (στοιχεῖα), and a fourth principle that is external and is not an element, and these four are the same by analogy for everything. The elements are form, privation and matter (τὸ εἶδος καὶ ἡ στέρησις καὶ ἡ ὕλη), and the principle that is not an element is the moving, or 'efficient,' cause (τὸ κινοῦν).[20] These four principles are the same for each kind, and as we move out of a kind and make cross-kind comparisons, they are still the same, but now the same only by analogy.

All of the items within a kind have the same three elements, but the elements in one kind are different from the elements in a different kind. Not only are the elements of sensible bodies (natural substances) different from the elements of non-substance kinds, but non-substantial kinds themselves also have different elements. To give examples of non-substances in different kinds that have as their elements different trios of matter, form and privation he lists:

(1) in [the kind] *color*:
pale, dark, surface;
(2) and light, darkness and air,
the things from which *day* and *night* are.[21]

Colors are attributes, and do not have matter in the way in which sensible bodies do. However, there is a way in which they do have matter, namely by analogy. For a color, *surface* is said to be its matter in this second,

[19] *Metaph.* Λ4, 1070a31–33; cf. 1070b16 ff.
[20] *Metaph.* Λ4, 1070b17–19, 22–23.
[21] *Metaph.* Λ4, 1079b20–21.

analogical way. The matter of a sensible substance stands to a substance as surface stands to a color, and this is what counts as the matter of a substance and the matter of a color being the same by analogy.

In fact, in a way the matter-analogue is matter, as Aristotle explains in *Generation and Corruption* 1.4:

> And while matter is most of all and strictly speaking that underlying thing which is receptive of generation and perishing, in a certain way the things underlying the other changes are matter too, because all of them are receptive of oppositions of some sort.[22]

This passage characterizes the role that is the same for matter and the various matter-analogues in terms of two factors: (i) it is what underlies (τὸ ὑποκείμενον), and (ii) what it underlies is an opposition (ἐναντίωσις). Speaking quite generally, the role of matter is to be what *underlies* an *opposition*. An opposition is a pair of opposites (ἐναντία). Although it is not possible for an underlier to possess both opposites at the same time, the underlier is the kind of subject that is *receptive* of both. This does not mean that it can have either one in any and all conditions. Rather, it is a subject that is of such a nature as to have either one of the pair, and is so even when with respect to one of them there are factors, either internal or external, that prevent the underlier from having it. For instance, Socrates is the kind of subject that is receptive of standing, and is so even when he is prevented from standing up because he is tied to a chair.

The passage in Λ4 also mentions change (μεταβολή). For any change, there is something that underlies a pair of opposites. Matter is what is receptive of generation and perishing, or else is the underlier for some other kind of change. However, although there is a difference between having an opposite and changing in such a way as to have it, what underlies a change from one opposite to another is always what underlies the opposition that includes both. At the start, the underlier must be receptive of both opposites, but have just one. It is impossible for a subject to possess both members of an opposition at exactly the same time. However, the subject that has just one of the pair, but is receptive of both, must be of such a nature as to receive the other member of the opposition. As a result of such a change it will have that opposite, and when that happens it no longer has the other member of the pair, the one that it had at the start.

22 *GC* 1.4, 320a2–5.

The role of a 'matter-analogue' for attributes

What more can be said about this general and analogical way in which both substances and their attributes have matter? The specific role played by matter in substantial generation does not have a counterpart in these other cases. For such a generable substance, the matter is what becomes the substance. It comes to be out of opposites (τὰ ἐναντία), where the opposites do not change.[23] This specific role is not general enough to include matter for attributes. After all, colors are opposites that do not change, and so they could not have matter for change as a substance does. No cause of color could play exactly that role, since a color is not the kind of thing that comes to be because some underlying matter becomes that color.

However, embedded in the causal role for each and every type of case, both for various substances and their various attributes, there is a triadic causal structure of matter, privation and form. The 'matter' of a color occupies a position in that structure that corresponds to the position occupied by the matter of a substance. That is, in the triadic causal structure of *surface*, *white*, and *dark*, surface occupies a position that corresponds to that occupied by matter in the triadic structure for generable substance. For matter to be what changes into a *substance*, the matter must (i) be what underlies opposites. Additionally, for it to be what underlies opposites it must (ii) be what is capable of receiving them both.[24] Described in this way, this is a role that is common to substances and their natural attributes. At the most general level, in each case one opposite is a form, and the other opposite is a privation of that form, and there is an item that occupies the remaining position in this type of structure:

> ELEMENT STRUCTURE:[25]
>
> <(1) what underlies: {(2) form, (3) privation}>

To enter into position (1) something must be the kind of subject that is capable of receiving a range of properties belonging to a kind, where the range is determined by the occupants of positions (2) and (3), the extremes that serve as the limits of the range. In the case of <surface: {pale, dark}>, the range determined by the opposites will include all and only the members of the kind color, and the intermediates are defined in terms

[23] *Metaph.* Λ2, 1069b3–7. Although the substance that comes to be is not an opposite, that substantial form into which it changes is opposite to the corresponding privation.

[24] *Metaph.* Λ2, 1069b6, b14–15.

[25] This label is chosen to indicate that these three items are the principles that *Metaph.* Λ4 refers to as 'elements [στοιχεῖα]'; this terminology of 'elements' is used in a similar way in *Phys.* 1.6, 189b16–29.

of ratios of pale and dark. The properties from this range are modifications of *surface*, and this is of course the item in position (1). Hence although strictly speaking colors do not have *matter*, for a color there is a cause that is analogous to the matter of a generable substance. Colors are the various modifications of this kind of subject, surface.[26] Accordingly, properties in the range marked out for that genus will be certain conditions that a surface of a physical object can be in. Such properties cannot be what they are independently of *surface*, the position (1) item. A privation is a condition that something naturally suited for a form is in when the form is absent in some way or respect. Like form, it too is something that either is or is not, but does not come to be or pass away.[27]

Hence, even though a surface is not a subject of *change*, and a surface could not be something that turns into light, dark and the rest, it too must be a subject that is receptive of the opposites. It is in this way that surface is by analogy the matter for color. That is, surface is an underlying subject that is receptive of the opposites light and dark, where light is the form and dark the privation.[28]

When *Metaphysics* H4 assigns eclipses and sleep to the class of things that do not even have matter, and presents instead what I will call its 'matter-analogue,' this latter item is what plays this analogous role.

The matter for sleep does not have its causal role all on its own, independently of the role played by the other causes of sleep. Rather it has the role as part of a triadic causal structure:

> TRIAD-1:
> <an <u>affection</u> of the <u>underlier</u> due to the <u>agent</u>>

To emphasize these three causal factors, one could represent the structure as:

> TRIAD-2:
> <AFFECTION, UNDERLIER, AGENT>

For eclipses and for sleep, the first item, the AFFECTION, is a privation of some sort; and the second item, the UNDERLIER, is the subject so affected; and the AGENT is the nearest cause responsible for that subject

26 Additionally, for Aristotle surface is mentioned in the definition of the kind to which they belong.

27 There are various ways in which a subject may be deprived of form, and not simply the condition that is maximally opposed to possession of form.

28 When *Metaphysics* H5 uses the example of the coming to be of a pale man to illustrate the thesis that forms are or are not without coming to be or passing away he does not bring in the distinction between a physical substance and its surface. The example could easily be expanded to accommodate the distinction, but this kind of elaboration is not needed for the purposes of that chapter.

being affected in just that way. However, in each case the privation is only temporary, and is itself preceded by, and eventually will be succeeded by, a form or positive condition of that subject. When viewed more generally, the AFFECTION will be one of the properties from the range determined by an opposition of form and privation, and the UNDERLIER will be a subject of such a nature as to be capable of receiving the relevant positive and privative conditions. The third item, the AGENT, is what is also referred to as the *efficient cause* that acts on the UNDERLIER in such a way that it comes to possess the relevant AFFECTION. According to *Metaphysics* Λ4, this agent, or efficient cause, is a principle that is not an element, and hence:

> while analogically there are three elements, there are four causes and principles; but the elements are different in different things, and the primary cause as that which effects change is different for different things.[29]

To sum up, the underlier may be called 'matter,' and it is in an analogical way that attributes and properties are said to have matter. For each such underlier, there is also a cause that is analogous to the efficient cause of a generated substance. This agent-analogue is what is capable of affecting the underlier in such a way that it acquires an affection that plays a role analogous to the form, or the privation, involved in a substantial change.

The matter-analogue for eclipses

Let us now look at the two cases discussed at the end of *Metaphysics* H4, eclipses and sleep, in a bit more detail. The way in which Aristotle discusses the matter-analogue in each relates the 'matter' to the other two items listed in TRIAD-2. The relevant discussion starts with the sentence that denies that they have matter: "Nor indeed is there matter for those things that exist by nature but are not substances, instead 'that which underlies' [τὸ ὑποκείμενον] is substance."[30] In both the case of a natural substance *and* the case of a natural attribute there is something that underlies, but whereas in the former case what underlies is *matter*, he tells us now that in the latter it is a *substance*. One of the key texts that draws the relevant distinction between two ways of underlying is a passage from the introductory remarks of *Metaphysics* Z13. This passage informs us that we have already discussed (in *Metaphysics* Z):

[29] *Metaph.* Λ4, 1070b25–27. [30] *Metaph.* H4, 1044b8–9.

> 'what is underlying' [saying] that it underlies in two ways, either (i) being 'some this', just as the animal [underlies] its modifications, or (ii) as the matter [underlies] its actuality.[31]

The subject that underlies in the *first* way is described as τόδε τι ὄν, being 'some this.' This is a substance, such as the moon, or a human. For instance, the subject that underlies musicality is a human, which is a primary substance, and is characterized by Aristotle as 'some this.' Likewise, the substance that underlies a privation of light (στέρησις φωτός) is the moon, and this is also a particular, primary substance.

However, what underlies in the *second* way specified in Z13 is not some substance (nor is it 'some this'), but rather is called in this text the '*matter*.' Indeed, what is a subject in this second way is the matter for the various items that are the subjects in the first way – those subjects that underlie attributes, and are described as 'some this.' These subjects, at least those that are *generable* substances, are themselves composites of matter and form.[32] However, neither an eclipse nor sleep is a composite of matter and form, where the matter underlies the form in this second way of underlying.

In causal inquiry about an eclipse, instead of looking for its matter, one should look for a *substance* that underlies something. Accordingly:

> For instance, what is the cause of an eclipse, what is the matter? For there is not [matter], rather[33] the moon is what undergoes [it] [τὸ πάσχον].[34]

This is meant to contrast with the case of generable substances, such as a human. In one of four ways in which something is a cause, a generable substance *does* have matter as one of its causes. However, eclipses do not have matter, but one of the causes of an eclipse is a substance, the moon. That is, for an eclipse, what underlies is not its matter, but rather the moon. Nonetheless, in both cases, that of an eclipse and that of a human, one of the causes *is something that underlies*. The difference that he stresses is that for an eclipse this is a substance (and not matter), whereas for a human this is matter (and not a substance).

There are analogues to at least two of the other causes for a generable substance. In the eclipse case, the form-analogue (the cause ὡς εἶδος) is what he calls the 'account' (ὁ λόγος), and this is identified as a privation

[31] *Metaph.* Z13, 1038b4–6.

[32] See, for instance, *Metaph.* Θ7, 1049a29–30, which singles out a man as what underlies modifications, and explicates this as "body and soul."

[33] Reading ἀλλ' ἡ with E, instead of ἡ (J, Ab).

[34] *Metaph.* H4, 1044b9–11.

of light.[35] Such an account, however, is said to be *unclear* (ἄδηλος) unless it is accompanied by the efficient cause:

> And [the cause] 'as form' is the account, but the account is unclear unless the [efficient] cause is added. For instance: what is an eclipse? Privation of light. And if one were to add 'by the earth coming in the middle,' this would be the account with the [efficient] cause.[36]

This passage first tells us that the earth is the efficient cause that destroys the light,[37] and subsequently says that the account *with* the efficient cause is 'deprivation of light *by the earth coming in the middle*.'[38] Given his injunction to state all of the possible causes, the full statement of causes should in that case be something like 'the moon's deprivation of light by the earth coming in the middle (of the moon and the sun).'[39]

Why is the account unclear without the efficient cause? At least part of the reason is simply that without the efficient cause such an account is *too general* to be something distinctive of just lunar eclipses. There are various ways in which something – even the moon – could be deprived of light. To be sure, an eclipse is some sort of deprivation of light (στέρησίς τις φωτός),[40] but saying this alone does not specify the sort of deprivation that it is, at least not in a way that would distinguish it from other ways of being deprived of light. Aristotle's proposal is not that this is clarified by adding some more determinate specification of observable features of the moon's light deprivation, or a fuller phenomenological description of what eclipses look like to observers. The proposal is rather that this (allegedly unclear) *logos* is made clear by accompanying it with a statement of the efficient cause that is responsible for the subject, the moon, having that property, or form-analogue.

[35] Note that in *Metaph.* Λ4, 1070b16–21, light is a form, darkness its privation, and air the matter. In H4, for an eclipse the form-analogue is a privation. In the case of a substance, he uses the word '*logos*' for the account or definition that expresses its form, and sometimes uses it for the form it expresses rather than the linguistic item itself. His use of the term '*logos*' has a similar duality of use in connection with definitions of natural phenomena, although in the two examples of eclipse and sleep the causal factor corresponding to a substantial form is not a form, but rather a privation. This general characterization of στέρησις needs to be worked out in different ways for different circumstances. For instance, *Metaph.* Θ1, 1046a31–35 distinguishes various ways in which something is called στέρησις. Nonetheless, in H4 he is speaking at a high level of generality about causes of natural phenomena, and drawing finer distinctions that do not generalize is not needed for this purpose.

[36] *Metaph.* H4, 1044b12–15.

[37] 1044b10–11. The light has as its subject the moon, and when its light is destroyed that subject is characterized by a privation of light.

[38] 1044b14–15.

[39] There is likely no final cause for an eclipse; see n. 43 below.

[40] One should compare this with *Post. An.* II.8, 93a23.

The mention of the efficient cause of an eclipse at 1044b14 is in fact the *second* mention of such a cause in the relatively brief and condensed passage at H4, b9–15. The *first* mention of the efficient cause occurs earlier in the passage:

> What is the cause in the sense of what effects change, i.e. extinguishes the light? The earth.[41]

This identifies the earth's causal role solely in terms of its causal *effect* on the item he has already listed as the subject for an eclipse, the moon. An eclipse occurs when the earth destroys the moon's light. Although this way of putting things brings in the subject (moon), and an attribute of the moon (its light having been destroyed), and an efficient cause for the subject having that attribute, it is not yet what he takes to be the 'clear' statement of either the form-analogue or the efficient cause. It tells us that a particular substance, the earth, is the efficient cause of the eclipse, but does not indicate what it is about the earth that is responsible for the effect. The earth is not the efficient cause simply in virtue of being the earth, or even simply in virtue of it being whatever it is that extinguishes the light. It is because the earth comes to be 'in the middle,' and thus blocks the light of the sun, that the moon is affected in such a way as to be deprived of light. To state the nearest efficient cause it is not sufficient simply to cite the earth itself as what brings about the relevant effect, since the earth does not cause this at just any time or in just any condition, but rather only when it comes to be in the middle.

With this fuller statement of the efficient cause, a statement involving the condition it is in when it produces the relevant effect, Aristotle takes it that one is now in a position to give a clear statement of the form-analogue as well. The form-analogue is not made clear simply by mentioning the earth as what destroys the light, but rather by mentioning what condition the earth is in such that by being in just that condition, something else (the moon) is deprived of its light. Both the efficient cause and the form-analogue are unclear if the efficient cause is identified simply as the earth, or as the earth destroying light. However, both the efficient cause and the form-analogue are clear once one knows what condition the earth is in when it brings about the light deprivation.

If this is the case, then the form-analogue stated *together with* the efficient cause is clear. The sort of privation of light that the moon undergoes in an eclipse is that sort of light deprivation that comes about by the earth being

[41] *Metaph.* H4, 1044b11–12.

in that condition (being in the middle). This efficient cause (the earth's coming in the middle) is the cause of the privation of light,[42] and although the form-analogue and the subject for the form-analogue, the moon, are also causes, this efficient cause is here singled out as 'the cause.' Citing this cause is what explains why the former item (the *pathos*) belongs to the latter item (subject, the moon).

Of course, even though the earth is cited as the efficient cause that brings about the eclipse, the earth itself does not actually *do* anything to put itself in the middle. After all, for Aristotle, the earth is itself stationary, and the moon and the sun are the objects that move around it. On occasion, as a result of what they are doing, they move into a configuration with the earth directly between them, specifically in just the way it is positioned when it blocks the moon's source of light. The motions of these other things themselves have efficient causes, and these are remote causes of the eclipse. Although they are not the 'nearest' causes, they are certainly part of the larger causal story of eclipses. A scientist who knows these causes as well has a fuller scientific understanding of the causal account of how and why eclipses occur. They are not, though, the causes with which the H4 passage is directly concerned, and they do not turn up in its statement of the causes of an eclipse. For this purpose he enjoins one to state only those causes that are nearest.[43]

By stating nearest causes one invokes both the matter-analogue (what is acted upon, or affected) and the efficient cause-analogue (what so acts, or the agent). Citing the causes in this way makes the *logos* clear not only by making it more specific, but also by making it evident that the relevant modification (the *pathos*) just is the condition the subject is in as a result of what the agent, the efficient cause, does. Whereas the moon is something that is affected, it is the earth, the efficient cause, that affects it in such a way that it is deprived of its light. *What the latter is doing* insofar as it acts (and does so in the way in which it is an efficient cause for X), and *what the former is undergoing* insofar as it is being affected (and is affected in the way in which it is in this other way a material cause for X), are the same. They are the same in whatever way the action of an agent is quite generally the same

42 1044b13, 15.

43 Also, it might turn out that there is no single efficient cause for the earth's coming to be in the middle, but rather a cause for the moon being where it is and another cause for the sun being where it is. If there is an efficient cause for them simultaneously being in those positions then there is such a unified cause for the earth's coming to be in the center. Probably there is no final cause for an eclipse (1044b12). But there are final and efficient causes specifically for the motions of the moon and the sun, and as parts of a more comprehensive teleological system *perhaps* also a final/efficient cause for the way the motions are coordinated.

as whatever it is that the patient undergoes. Of course, Aristotle also holds that there is a way in which what it is for X to be acting on something Y is not the same as what it is for Y to be affected by something X. These are not the same 'in being.' However, he holds that the *account* of the passive capacity of the patient to be changed by another must contain the account of the corresponding active capacity of the agent.[44]

In the case of the eclipse, the earth need not be seen as an agent in just the way in which, for instance, a father is an efficient cause of a child. Rather, it is the analogue of the efficient cause of the generation of a substance, and its agency is somewhat attenuated. (As already noted, it does not do anything at all to get in the middle.) However, by being in just the right position to block the moon's source of light it is what is directly responsible for there being an actually transparent medium for ordinary visual perception continuous[45] with the surface of the moon and any observers on earth. In fact, both the surface of the earth that faces the surface of the moon, as well as the part of the moon's surface that it faces, are equally deprived of light due to the position of the earth. The moon's loss of light during an eclipse is simply the condition of the transparent medium surrounding it that results from the earth's blocking its source of light.

The matter-analogue for sleep in recent scholarship

According to the H4 text, sleep is also in a class of natural things that lack matter. It too exists by nature, but is not a substance. Since sleep is not a substance, it does not have matter for generation and perishing. Nonetheless, when natural things and their causes are treated at the level of generality commensurate with that class, the cause 'as matter' (ὡς ὕλη) is one of their causes in each and every case. Accordingly, to inquire into, or to state, the causes of sleep correctly one must identify the cause that is related to sleep as matter is related to the generable substance of which it is the matter. Sleep does not have matter in the established sense of the term, but does have a subject that plays a role corresponding to the role of matter for generable substances. Furthermore, just as in the case of an eclipse, the matter-analogue does not play its role as the kind of cause it is except in concert with analogues to the cause 'as form' and the cause 'as that which effects change.' Hence in the case of sleep its 'matter,' or matter-analogue, has its role as part of the unified, tripartite causal structure in which it figures as the cause 'as matter.'

[44] See *Metaph.* Θ1, 1046a9–16. [45] Or at least contiguous.

Accordingly, one might expect it should not be too difficult to determine what the matter, or matter-analogue, for sleep is supposed to be. After all, his *de Somno* proposes to investigate, among other things, the cause due to which both sleep and wakefulness belong to animals,[46] and in the course of this investigation explicitly brings up his doctrine of the four ways in which there are causes.[47] Additionally, the concluding remarks of that work claim that it has been stated what the cause of sleep is.[48]

Aristotle's doctrine of four different ways in which things are said to be causes provides a crucial part of the overall framework for the investigation of the causes of things that exist by nature. The *de Somno* invokes this doctrine halfway through *Somn.* 2 in the following way:

> We must state through which cause sleeping occurs, and what kind of affection it is. And since there are several types of cause (for we say that (i) that which is for the sake of something, and (ii) that from which the origin of the change is, and (iii) the matter, and (iv) the account [λόγος], are [each] a cause), first we say that since nature acts for the sake of something, and this is some good . . . [49]

Just as in the *Metaphysics* H4 passage, though with slightly different vocabulary, he here lists the 'final' cause, the 'efficient' cause, the 'material' cause and the 'formal' cause, respectively. The quoted sentence goes on to argue that sleep is for the sake of the preservation (σωτηρία) of animals, and that waking is an end (τέλος). Their perceiving and their being sensible are, for an animal to which either belongs, not simply *an* end or a good, but rather *the good* and the best. We are here told that sleep has a final cause, and that this is a certain kind of preservation of an animal's nature.

If it is accepted that there is such a final cause, what more can be said about whatever other types of cause there are for sleep, and how they relate to the final cause? Among other things, there should be an efficient cause that originates a change, the result of which is that something has the 'form' of sleep (or whatever it is that plays the role of the item traditionally spoken of as the 'formal cause'). When that subject has been changed by the efficient cause in that way, with the result that the animal itself is asleep, this is itself in aid of some good for the animal. The result is a condition of rest that *preserves* the natural perceptual capacity, the natural capacity that the animal actively uses during its periods of wakefulness. Arguably, the connection with a final cause is not simply that what the efficient cause

[46] *Somn.* 1, 453b14; see also 2, 455b13. [47] *Somn.* 2, 455b14–16.
[48] *Somn.* 3, 458a25–26. [49] *Somn.* 2, 455b13–18.

brings about is this good (preservation), but more precisely that it acts *for the sake of* living. The 'formal cause' that is brought about is, when brought about in the relevant part of the animal and at the times at which rest is both needed and beneficial, not simply for the sake of preservation, but ultimately for the sake of the active life for which the animal's nature is the principle.

Nonetheless, the subsequent causal inquiry in the remainder of the *de Somno* does not, at least not explicitly, discuss the efficient cause in these terms. What about the material cause? Since the way an efficient cause is responsible for sleep is that it is an agent, there must also be a patient. This would be whatever it is that is able to be affected in such a way as to have the 'form' (or analogue of form) specified in the *logos* of sleep. This would be the cause for sleep that corresponds to the material cause of a substance. However, not only is this not what scholars have looked for in the *de Somno* when considering the material cause of sleep, but there is not even consensus as to which of these four types of cause that work takes to pertain to the explanation of sleep.

Some scholars maintain that Aristotle employs all four causes in that work, but others argue that sleep lacks at least one of them. To further complicate matters, there is also disagreement as to whether the treatise handed down to us combines texts from *different* stages of his philosophical development. Although there is some kind of scholarly consensus that at least some part of the treatise does assign sleep a material cause of some sort, there is disagreement as to whether those passages in fact represent only an earlier view, and are in conflict with the theory developed in other parts of that work. Finally, there is not simply disagreement as to which causal roles are invoked in his explanation of sleep, but still further disagreement as to which factors play which causal roles. Even among those who treat the *de Somno* as a single work expressing a single theory, there is considerable disagreement as to how matter fits into the theory.

These disputes are in part connected with worries about final causality. If we return to *Metaphysics* H4 we can see that although it does ask about the 'cause as that for the sake of which' (ὡς οὗ ἕνεκα), or final cause, for two of the cases discussed earlier in the passage, before the topic of sleep is addressed, it says nothing at all about finality in connection with sleep. This silence has been taken as evidence that at the time he wrote this passage he had abandoned the view that there is a final cause of sleep.

In a moment I will consider a view according to which this provides confirmation for interpreting the *de Somno* theory along these lines. First, however, it should be stressed that this feature of the H4 passage does not by itself show that at the time Aristotle wrote it he thought that sleep lacked a final cause. After all, the discussion in H4 of eternal natural substances did not even bring up the topic of final causality, but that is not a reason to think that they altogether lack such a cause. If it were, it would also be a reason to think that they did not have efficient causes or formal causes, since those are not mentioned either. Additionally, the discussion of principles in *Metaphysics* Λ4 omits final causality (as does Λ1–5 as a whole), but once again, that is no reason to think that when he wrote it he rejected, or perhaps had not yet formulated, the view that natural substances have final causes.

That said, Malcolm Lowe has taken H4's omission of a final cause of sleep as evidence that at the time it was written Aristotle had abandoned teleological explanation for such natural attributes. He has argued that this text in the *Metaphysics* represents a late stage in the development of Aristotle's thought about causality, and says of this text: "it is implied, sleep is not something having four causes: there is no material cause; the final cause may not exist; the formal cause is given by the definition, but this is incomplete without the efficient cause . . . "[50] Lowe holds that the passage not only shows that sleep lacks a material cause, given that 'the formal cause' is not a form, but instead a 'privation of sensation,' but also that the final cause cannot be identified with the formal cause. He takes this to go against what he describes as Aristotle's "well-known tendency" to see the formal and final cause as somehow identical.[51] Furthermore, since the formal cause (i.e., the privation) "is unclear if it is without the [efficient] cause,"[52] Lowe proposes that Aristotle's view is that, in light of this, sleep cannot be defined without a specification of the efficient cause.[53] Hence:

> the formal cause amounts to the efficient cause plus something else. Thus, instead of the well-known four causes, the reader is offered (so to speak) merely one-and-a-half.[54]

[50] Lowe 1978, p. 287. Lowe's conclusions on this matter are accepted by Gallop 1991; on p. 130, speaking of *Somn.* 3, he says "there is no indication in the text that a 'material cause' for sleep is being identified." Their joint view is effectively criticized by Everson 2007.

[51] Lowe 1978, p. 286. Lowe does not indicate why, given that sleep is a privation, the final cause could not be the correlative form. For further criticism of this see Everson 2007, pp. 504–7.

[52] Lowe 1978, p. 287.

[53] Against this, it should be pointed out that Aristotle does not suggest that the *logos* for the form-analogue is incomplete, or requires revision. His point is rather that the *logos* is unclear unless accompanied by the efficient cause, not that the *logos* needs to be expanded so as to include a reference to the efficient cause.

[54] Lowe 1978, p. 286.

On the interpretation that Lowe puts forward, this is not only the view of the H4 passage, but is also the view held when Aristotle wrote the bulk of the *de Somno*. Lowe argues that that work contains material from two different treatises on sleep, each of which represents a different stage in Aristotle's philosophical development, and that the later and better view that informs the majority of that work is the one that drops the four-cause model in favor of the 'one-and-a-half' model.

Lowe argues that a fragment from an earlier view was inserted in Chapter 2, at 455b13–34, and that a comparison of this fragment with the main treatise "reveals a significant development . . . in Aristotle's well-known theory of the four causes."[55] The fragment "boldly starts upon an investigation of four causes of sleep."[56] However, on his view, the main, and later, treatise says *nothing whatsoever* about the material and the final cause, and furthermore it too adopts the view that the formal cause cannot be defined without specifying the efficient cause. For Lowe, the H4 passage we have been considering provides independent confirmation that the theory of four causes was later replaced with the view that sleep has no material cause, and may not have a final cause.

My aim in the present chapter is not to resolve all of the many issues concerning the unity of the *de Somno*, nor to present a full argument for a new interpretation of that work. Instead I will address the more limited, and preliminary, task of exploring more fully how that work could be interpreted were it in fact following the methodology presented in the *Metaphysics* H4 passage. In particular, I will sketch out a way to interpret the *de Somno* theory such that it employs the causal structure, TRIAD-2, presented in the *Metaphysics*, and does so in a way that is perfectly consistent with a more expansive causal perspective that embeds this structure within a teleological framework. This is sufficient to show that the H4 passage does not itself provide independent support for Lowe's interpretation of the *de Somno*. However, more work would need to be done to assess fully the merits of my proposal as an interpretation of the *de Somno*, and its consequences for issues about the unity of the treatise and questions about philosophical development. The pattern for stating causes advocated in the *Metaphysics* H4 passages gives to the matter-analogue a specifiable causal role, that of something affected by an efficient cause in such a way as to be

[55] Lowe 1978, p. 279. His argument for this is developed in opposition to the earlier interpretation in Lulofs 1947. According to Lulofs, the treatise combines an unfinished *de Somno* A, which is later and only partially preserved at 453b11–455b13, and a *de Somno* B that is preserved in its entirety at 455b13–458a32.

[56] Lowe 1978, p. 286.

characterized by the relevant *logos*. This puts constraints on what entities we could assign as the causes of sleep were we using this pattern in the *de Somno*, and does so in a way that gives us a basis for opening up discussion as to what its theory should be like if it does conform to that model.

Additionally, Aristotle's treatment of other psychological affections that are suitably similar can be approached in this way as well. Not only is there a comprehensive and detailed treatment of the causes of sleep in the *de Somno*,[57] but there is also an important methodological passage about attributes of this sort in *de Anima* I.1 that has been the locus of a great deal of scholarly debate.[58] Among other things, this passage discusses how affections common to body and soul ought to be defined, as well as considerations about stating both their matter and their form. This too is directly relevant to sleep, since Aristotle clearly considers sleep to be one of those affections.[59] The *de Anima* I.1 passage characterizes a fairly wide range of such psychological affections as "enmattered accounts" (λόγοι ἔνυλοί),[60] and tells us that the natural scientist defines them in a way that involves both their matter and their form. Among other things, the passage first motivates his claim that they are 'enmattered accounts,' and subsequently gives some guidance as to the general way in which the definitions of these affections of soul ought to be structured. To illustrate these points it briefly discusses the way in which a dialectician differs from a (non-Aristotelian) natural scientist when defining anger. The former defines only the form (e.g., 'desire for revenge,' or something like that) whereas the latter's definition gives just the matter (e.g., 'the boiling of the blood, or the hot, around the heart'). By way of contrast with both of these, the appropriate procedure for a natural scientist is to define anger in a way that combines both the matter and the form.[61]

Unfortunately, that text does not give this type of combined definition for anger, and scholars have not agreed on exactly which definitional methods or procedures he thinks a natural scientist ought to use for this purpose. Even though his remarks suggest candidates for the matter and form of anger, they do not say just what it is for something to be the matter, or the form, of an affection common to body and soul. Since he recommends applying hylomorphic analysis not just for substances and artifacts, but

[57] In *PA* II.7 there is a brief discussion of the role of the brain in the production of sleep (653a10–20).

[58] *De An.* I.1, 403a3–b19. The literature on this is enormous, and covers a wide range of topics. The reader can find an accessible discussion and interpretation of many of the central issues in Johansen 2012, Chapter 8, pp. 146–69. For an illuminating discussion of interpretive difficulties, and opposing views, concerning the way matter figures in definitions of these psychological affections see Charles 2009 and Caston 2009.

[59] *Somn.* I, 453b24–454a11. [60] *De An.* I.1, 403a24–25. [61] *De An.* I.1, 403a29–b9.

also for their attributes, additional guidance as to how such analysis is to be extended to these cases would be useful. The methodology given in the *Metaphysics* H4 passage gives such guidance, and invites comparison with the *de Anima* discussion. This is not the place, though, to attempt an interpretation of the *de Anima* passage. However, if one is armed with H4's account of the role assigned to a matter-analogue for sleep, this can then be used to put restrictions on the candidates for matter and form in the interpretation of other attributes common to body and soul. This can serve as a basis to explore the interpretations on offer of the *de Anima* passage from the vantage point of the methodology presented in the *Metaphysics.*

One cannot simply assume that the pattern specified in H4 is in fact what he employs, or intends to employ in other works. Nonetheless, this kind of comparison is useful in part because there is no scholarly consensus as to what, if anything, Aristotle takes the material cause of sleep to be, or how he thinks matter and form are related to each other in the causal inquiry into such attributes in general. There are, as we shall see, interpretations of his views on matter that do not treat the matter as the subject that is affected by the efficient cause. Exploring how they contrast with the methodology taken from H4 is a way to see how the conception of matter they employ differs from the one developed in this chapter, and how such alternatives relate to his hylomorphic approach to natural science.

As we shall see, the matter-analogue for sleep in the H4 passage is a bodily part that is affected by an efficient cause. In blooded creatures, he takes this to be the heart. The current interpretations of Aristotle's views on how sleep, or related attributes, have a material cause do not consider candidates such as the heart for just this role. Accordingly, before considering the passage in any more detail I will first give a brief, and somewhat selective, overview of a few of the views about the material cause of sleep in the *de Somno.* There are quite a few textual issues that a full treatment would require, but for this kind of overview they have to be set aside.

The most well-known interpretation is that of W. D. Ross. He writes that Aristotle: "offers no separate account of the material cause, but it has become plain that this is the hot matter contained in food."[62] By way of sharp contrast, Robert Bolton argues that the material cause is the "seizing up of the primary sense organ, the heart," and that this material cause is itself caused by the efficient cause, which he identifies as "the necessary

[62] Ross 1955, p. 44. He further states that "[chapter 3] makes it plain that the material cause is the hot matter contained in food" (457a33–b1) (Ross 1955, p. 260).

massive return of cooled material to the region of the heart."[63] Both of these views in turn contrast with the claim by Stephen Everson that the material cause is the inward concentration of the hot matter,[64] and that for Aristotle "explanation of sleep in terms of its final and material causes is entirely sufficient."[65] For him, the material causes are material changes, and sleep neither has nor requires efficient or formal causes.

As different as the views of these three authors are, one thing they have in common is that they do not identify the material cause with the subject that is affected in such a way as to possess the privative modification, or form-analogue. Indeed this is what one finds quite generally throughout the literature on the topic. On any interpretation of the *de Somno* theory a blooded animal is asleep when its heart is affected in such a way that it is (temporarily) incapable of the kind of perceptual activity characteristic of its waking life. However, this has not led scholars to identify the heart with the material cause. But, if something other than the heart is playing the role of the material cause, then it is not part of the material cause's role that it is what is perceptually incapacitated due to the activity of the efficient cause. If that is correct, then one needs not just an alternative *candidate* for the material cause, but also an alternative account of the role itself that differs from what is assigned to it in *Metaphysics* H4. It is now time to consider a few more details of that text, and how they relate to the explanation of sleep.

The matter-analogue in *Metaphysics* H4

When *Metaphysics* H4 turns to a consideration of the causes of sleep, Aristotle notes a respect in which there is a difference between the case of sleep and the earlier case of an eclipse:

> But for sleep it is unclear what the primary thing affected [τὸ πρῶτον πάσχον] is. But is it not the animal? Yes, but [it is] this in respect of what, i.e. what primary thing [κατὰ τί, καὶ τί πρῶτον]? The heart or something else.[66]

To be sure, it is the animal as a whole that sleeps, and when it does so it has sleep as one of its attributes. However, the substance (the animal as whole) is affected in such a way that it is asleep *in respect of some primary thing*. This

[63] Bolton 1997, pp. 104–5. This brings an interpretation of the material cause of sleep into line with an interpretation of the material cause of anger in *de An.* 1.1 as the boiling of blood, or warm substance, around the heart.

[64] Everson 2007, p. 513, n. 35.

[65] Everson 2007, p. 519.

[66] *Metaph.* H4, 1044b15–17.

primary thing is not the whole animal, but rather a non-uniform, material part of the animal, such as the heart, or something else of that sort. When one asks *in respect of what* (κατὰ τί) the animal sleeps, this could be asking either for the *form* of sleep or for the primary, *proximate subject* for that form,[67] where the former is predicated of the latter. What he has in mind here is the proximate subject for the attribute, and this is not the animal itself, but rather a functional part of the animal, such as the heart, or some other such part.

In this passage, what is predicated of this primary subject is called the modification (τὸ πάθος). This is a modification, or affection, of the heart, and the animal as a whole is asleep in virtue of its heart being affected in just that way. The relevant way is an immobility of some sort (ἀκινησία τοιαδί).[68] That is, sleep occurs just in case an animal as a whole is asleep, and this takes place when, and because, the animal's heart (or some other such organ, if it is not the heart) is *immobilized in a certain way.*

Accordingly, this methodological procedure involves an extra step, one that is not present in the procedure he specified for seeking the causes of an eclipse. He did not specify some material part of the moon that is the primary subject that undergoes modification when an eclipse occurs, such that the moon as a whole undergoes eclipse because this part of the moon is modified in a certain way. In the eclipse case, the form-analogue (the cause ὡς εἶδος), or what he calls the 'account' (ὁ λόγος), was said to be privation of light.[69] This is the account that is supposed to be unclear (ἄδηλος) unless it is accompanied by the efficient cause.[70] The privation in the case of sleep is also unclear in very much the same way. Just as there are various ways of being deprived of light (not all of which are ways of being eclipsed), so too there are various ways in which an animal could be deprived of the ability to perceive (not all of which are ways of being asleep).

The animal is by nature a perceiver, and in its waking state it engages in perceptual activity. Additionally, it goes through regular periods of these states of rest during which it is unable to perceive. This is, though, not just any inability to perceive, but according to the *de Somno* is an inability that is for the sake of this good end (perceiving). Even if we were to bracket the

67 See *Metaph.* Δ18, 1022a14–19.

68 *Metaph.* H4, 1044b19. See also *Somn.* 1, 454b5–6 where he says that the affection is an incapacity (ἀδυναμία) due to an excess of being awake. At b10–11 it is described as "a certain affection of the perceptual part, a bond or inability (ἀκινησία)."

69 Note that in *Metaph.* Λ4, 1070b16–21, light is a form, darkness its privation, and air the matter. In H4, for an eclipse the form is a privation.

70 *Metaph.* H4, 1044b13.

teleology, inabilities due to "unconsciousness, choking of a certain sort, and fainting"[71] do not count as sleep. How, then, does Aristotle distinguish sleep from other cases of perceptual inability? In that work he goes about this by identifying the proximate efficient cause. If the work follows the *Metaphysics* H4 pattern, then this is the agent that affects the primary organ of sense in such a way as to bring about perceptual incapacity. Identifying this agent would make the account of sleep clear in the way that the efficient cause of a lunar eclipse makes clear the way in which the moon is deprived of light when eclipsed. As in the astronomical case, the activity of the efficient cause is the same as whatever it is that the patient is undergoing. That is, it is the same in whatever way it is true quite generally that what agents do and patients undergo are the same.

However, there is an additional unclarity, not present in the astronomical case, and that is that there is some part of the animal that is the primary thing affected by the efficient cause. Initially, prior to inquiry, it is not clear what this might be. In the *de Somno* the primary thing affected when an animal sleeps is the primary organ of its perceptive faculty,[72] whatever that may be. For blooded animals this is the heart, and for others the part analogous to this. To determine that the heart is the relevant organ the treatment in *Somn.* 456a4–6 first identifies it *via* its functional capacity, and then relies on something established elsewhere, that the organ for that functional capacity is the heart of a blooded animal. In the case of blooded animals quite generally the primary subject for the relevant *pathos* is the heart. However, in *Somn.* 2 it is first identified via its *functional role*. It is the master sense-organ (τὸ κύριον αἰσθητήριον).

In *Somn.* 2 Aristotle also argues that the part of the body in which sleep originates is the heart. Since the heart plays this functional role, it is the subject that is affected. The argument at the end of *Somn.* 2 for the claim that sleep originates in the heart is the conclusion of a phase of inquiry that started at the beginning of that chapter with the question as to why an animal sleeps or is awake, and to what sense or senses this is due.[73] This provides his grounds for holding that sleep originates in the master organ for perception, which is the heart in the case of blooded animals, and the analogous part in bloodless animals, including insects. It is the heart

71 *Somn.* 3, 456b9–11; cf. 2, 455b2–10.

72 *Somn.* 2, 455a21. This is the sense-organ that controls all of the other sense-organs (a33 ff), and is also characterized as the common faculty that accompanies the others, by virtue of which one perceives that one is seeing, or hearing (a15–17). Hence, it is a part that is common to all of the sense-organs (a19–20). He also refers to it as the primary sense-organ (τὸ πρῶτον αἰσθητήριον) (458a28).

73 *Somn.* 2, 455a4–5.

(rather than, for instance, the brain) that is the origin of the changes that result in the kind of perceptual incapacity characteristic of sleep.[74] Hence the heart is both a remote efficient cause of sleep and also its proximate material cause.

Animals are self-regulating systems, and the normal transitions between wakefulness and sleep involve the condition of this physical organ alternating between periods in which it is affected in such a way that the animal sleeps and periods in which it is not. The efficient cause of sleep will be whatever it is that is responsible for the heart being affected in that way. However, the claim that the heart is the subject that is affected when the animal sleeps does not by itself tell us what the efficient cause is. Nonetheless, it does narrow down the list of candidates. Whatever the efficient cause is, it must be something that affects the heart.

Since the heart is the primary organ of perception, Aristotle looks for this kind of cause by considering what the animal does when it first starts its perceptual life. This would be when the heart starts performing its perceptual function, and occurs once the animal starts eating and growing.[75] That is, he claims that at the same time that the heart begins to function as an organ of perception the animal also starts nourishing itself by ingesting food. This is converted into an exhalation that enters its veins, and is transformed into blood. The blood is its 'ultimate' form of nutriment, and by traveling through the vascular system it goes to its 'origin,' the heart. Hence, on his view, once the perceptual system is operative, the primary organ of perception is not only affected in such a way that the animal perceives, but it also is the place to which the nutriment produced by eating travels. In some way still to be specified, this nutriment, in its partially digested form, leaves the area of the heart, moves to the brain, and once again returns to the heart. Upon its return, at least if the return is sufficiently massive, the heart is immobilized in such a way that it is temporarily incapable of being affected by external sense objects in such a way that the animal perceives them. Hence, the animal cycles back and forth between sleep and wakefulness, and the partially digested nutriment is the efficient cause that puts the heart in the condition for sleep. This efficient cause is intimately connected with the nutritive activity that commences with the animal's perceptive activity.

74 It is also the part of the body that originates motion. This occurs, he says, in the area intermediate between the head and the lower abdomen (*Somn.* 2, 455b34–456a14).

75 Of course there are ways in which the organism was nourished and grew before it started perceiving, but these do not involve eating food.

In the case of an eclipse the efficient cause is something that is external to the moon – it is neither the moon, nor a part of the moon. By way of contrast, for sleep the efficient cause turns out to be something that is not external to the animal itself. *Metaphysics* H4 itself does not identify this efficient cause, as it did for the eclipse. It does, though, require that one should ask for these causes of sleep:

(1) the primary or proximate subject [τὸ πρῶτον πάσχον]
(2) the efficient cause [ὑπὸ τίνος;]
(3) the modification [τὸ πάθος] of the proximate subject (i), and then answer the question:
(4) [it is] this [modification] in virtue of the primary thing undergoing what? [αὕτη τῷ τί πάσχειν τὸ πρῶτον].[76]

We have already seen that he identifies the first of these, the primary subject, with the heart or some such part. If this listing of causes in H4 follows the same pattern as the previous case, that of the eclipse, then the second item (that by the agency of which sleep occurs) would be something that is responsible for affecting this primary sense-organ in such a way that it has the third item, the relevant πάθος (the form-analogue of sleep), present to it. This would be another application of the causal structure:

TRIAD-2:
<AFFECTION, UNDERLIER, AGENT>

This passage does not here say what this agent might be, but rather asks in what way the primary subject (the heart, or its analogue in bloodless animals) is affected such that it is due to being affected in that way that it is immobilized. Given that what the patient has done to it is the same as what the agent does, one can determine the latter by determining the former, and the efficient cause will be whatever it is that does precisely that to the heart. Since the animal does not lose its perceptive faculty when it sleeps, the relevant inability must be due to the absence of one, or more, of the conditions required for external sense objects to affect the heart in such a way that the animal perceives them. If this is parallel to the case just discussed, that of an eclipse, the relevant condition would be that of the medium of perception, and the efficient cause would be whatever it is that is responsible for the absence of the appropriate medium.

What are the candidates in the *de Somno* for an efficient cause that would deprive the heart of its proper medium for perception? Since sleep occurs

[76] *Metaph.* H4, 1044b19–20.

when the corporeal stuff (σωματώδης) in the blood (or its analogue) descends from the brain to the heart, the chief candidates would be the brain itself, and the corporeal stuff it cools, thereby occasioning its return to the heart.[77] According to *Somn.* 3, sleep occurs when the exhalation from food eventually flows back upon the heart in a mass. The process is as follows. The exhalation from food initially enters the blood, and eventually makes its way to the region of the heart. After being heated there, it is first carried up by that heat to the brain, and then having been cooled descends downwards toward the heart where it is said to drive away 'the hot [ἀπώσῃ τὸ θερμόν].'[78] Scholars do not agree on just what 'the hot' is that gets driven away, or how driving it away is supposed to result in sleep. The controversy surrounding this point is mainly concerned with whether the heart is cooled or heated by the returning corporeal material, and what cooling, or heating, do to block occurrent perception.[79]

However, given that one is looking for an efficient cause that blocks perceptual functioning in a way parallel to the blocking of the moon's light, what is relevant is not how heating or cooling affects the heart, but rather how one or both of these brings about the relevant change in the medium. If sleep is induced by a change in the condition of the medium required for the primary organ to engage in active perception, then just as the moon cannot be seen unless there is a continuous[80] medium between it and the organ of sight of an observer, so too the external objects of sense perception cannot affect the primary organ of perception unless there is a continuous medium, in the right condition, between the primary organ of sense and the external sense objects.

77 In favor of the brain, the *Parts of Animals* tells us that this part "produces sleep in those animals that have one; while in those without one, it is the analogous part. For by cooling the flow of blood from the nourishment, or on account of certain other similar causes, it weighs the region down . . . and makes the heat recede downwards along with the blood. Because of this greater accumulation in the lower region it produces sleep" (II.7, 653a10–17; Lennox translation). However, although the brain is *an* efficient cause, it is not the nearest, most proximate one. Once the blood flows to the brain and is cooled, it reverses direction and heads back down toward the region of the heart. In this way, once the blood has been cooled, the brain has done its part and makes no further contribution to controlling the causal chain that terminates at the heart. The brain is not responsible for what happens to the cooled blood before it enters or after it leaves the brain, and is not what acts on the primary sense-organ. Alternatively, in favor of the corporeal stuff, according to the *de Somno*, the solid stuff that is in the blood and gets cooled in the brain subsequently returns to the heart, which is the primary sense-organ, and when it does so this solid matter is in a condition that causes the heart's immobilization.

78 *Somn.* 3, 456b27.

79 For a thorough account of the controversy and related textual issues see Wiesner 1978.

80 For this purpose, the distinction between continuity and contiguity is not relevant just so long as all contiguous bodies are in the right condition for perception.

For colors, for instance, to affect the primary sense-organ there must be an actually transparent medium that is continuous between the external object of sight and the eye, and from there all of the way through the pores leading from the eye to the brain, and then through the vessels leading to the heart. Since the optimal conditions for perception occur when the purer and colder blood is in the upper regions of the body and the more turbid blood is in the lower regions, the more these conditions are disrupted, the less optimally the primary sense-organ will function.[81] This is, in fact, what Aristotle thinks happens when nutriment is being digested in the blood. When the exhalation from food enters the blood, the purer and rarer blood gets mixed with the turbid and thicker blood. When blood in this mixed state comes down from the brain in a significant quantity, and displaces the blood in the region of the heart, it is the blood in this mixed state that surrounds the heart. This condition continues until it is fully digested and the animal awakens.

When this partially digested nutriment surrounds the heart, the medium surrounding the heart is not in the condition appropriate for occurrent perception. In fact, it is instead in a condition that is appropriate for the reception of various sensory images that have been stored in the sense-organs. These are conditions that are ideal instead for continuous dreams.[82] However, once the digestion is complete, and the fine blood has been separated from the thick, optimal conditions for occurrent perception are restored and the animal awakens.[83] Prior to waking, the corporeal stuff that has descended from the brain to the heart plays the same kind of role that the earth plays in the explanation of an eclipse. It is what is responsible for the condition of the medium in contact with the underlier not being suitable for occurrent perception. In each case perceptible qualities are thereby prevented from affecting the organ of perception. The main difference is that during eclipses the medium in contact with the sense object is unsuitable for perception, whereas during sleep it is the medium surrounding the heart that is unsuitable. In the latter case, it is this incompletely digested blood that pushes away the blood around the heart, and ends up surrounding the heart itself. Although both heating and cooling are involved in the physiological changes that result in this placement of the partially digested nutriment, neither they nor the brain that does the cooling are sufficient to shut down perception. The nearest, or proximate, cause of this is rather the mixed blood that has come to surround the heart. With respect to the explanation of sleep, the questions as to whether the

[81] See Gregoric 2007, pp. 44–45. [82] *Somn.* 3, 461a25–27. [83] *Somn.* 3, 458a10–25.

heart is cooled or warmed are no longer relevant.[84] Once the medium of perception is changed in such a way that external sense objects no longer are able to affect the primary sense-organ, neither heating nor cooling of the heart itself plays a further causal role in bringing about sleep, although both played crucial roles at earlier, more distant, stages.

Whether this should be accepted as an interpretation of the explanation of sleep in the *de Somno* depends in part on textual issues. I will conclude this section with a brief discussion of a few of these. If we were to follow Ross, and delete the word ὑπὸ from the manuscript reading ἡ ὑπὸ τοῦ σωματώδους, then at the end of *Somn.* 3 Aristotle is stating that the cause of sleep is "the ἀντιπερίστασις of the solid stuff,"[85] where the ἀντιπερίστασις would have to be the massive reverse flow of the partially digested blood from the brain to the heart. If so, the translation of the entire sentence would then be something like:

> It has been said what the cause of sleep is, that it is the massive reverse flow of the corporeal stuff, that is carried up by the natural heat, to the primary organ of perception.[86]

As already noted, Ross takes this corporeal stuff to be the *material cause* of sleep.[87] He is right to hold that this stuff is something material, and that it plays some causal role in the specification of the efficient cause in this passage. Nonetheless, the *de Somno* never calls it the material cause, and Ross never says what criteria for being a material cause it satisfies. Additionally, if it were the material cause, then the *de Somno* could not be using the causal pattern found in *Metaphysics* H4. The partially digested nutriment is obviously not the primary organ of sense, and it is not the part of the body that is affected when the animal as a whole is asleep.

Alternatively, if the partially digested nutriment that descends from the brain to the heart is what causes the relevant perceptual incapacity, and does so simply by surrounding the heart (and hence depriving the heart of the right kind of medium for occurrent perception), then the ἀντιπερίστασις could not be the reverse flow of the corporeal stuff. In that case it would rather be the encircling compression itself. This is the meaning given by Liddell and Scott for the use of the term in this passage ('surrounding so as to compress'), and is also how Aristotle uses it in this text from the *Meteorology*:

84 Wiesner 1978, pp. 263–67, argues that an explanation in terms of the mixed state of the blood is not only not present in the *Parts of Animals*' explanation of sleep, but is in fact ruled out by the remark at II.2, 648b1–10 that the hot and cold are causes of sleeping. However, although there are differences between the explanations in the *PA* and in *Somn.*, the accounts may nonetheless at least be consistent since heating and cooling are not the *proximate* causes of sleep.

85 *Somn.* 3, 458a25–28.

86 *Somn.* 3, 458a25–28.

87 Ross 1955, p. 260.

> We can see that a surrounding compression [ἀντιπερίστασις] occurs for the hot and cold on each other. Hence in warm weather the lower parts of the earth are cold and in a frost they are warm. The same thing, we must suppose, happens in the upper region, so that in the warmer seasons the cold is concentrated by the surrounding heat and causes the cloud to go over into water suddenly.[88]

Given this understanding of ἀντιπερίστασις, when the mixed blood, after reaching the brain and being cooled, goes back down in a condensed mass to the heart, it is the surrounding compression by blood in the mixed state that is the efficient cause of sleep. This is the very condition that the undigested blood is in when the heart is affected in such a way as to be unable to perceive. In that case, we should retain the ὑπὸ at 458a25, and translate something like this:

> It has been said what the cause of sleep is, that it is the massive surrounding compression by the corporeal stuff – that is carried up by the natural heat – upon the primary organ of perception.[89]

Accordingly, just as in the case of an eclipse, the *logos* of sleep too is unclear until accompanied by the efficient cause. Sleep is not "any and every incapacitation of the perceptual part" (456b17–18), just as an eclipse is not just any and every loss of light by the moon. Sleep is rather the incapacitation of the perceptual part that is due to the descended nutriment, the mixed blood that is not yet completely digested, being in the condition of surrounding the heart in such a way as to render it incapable of perceiving. This is parallel to the moon's loss of light due to the earth being in between the moon and its source of light in such a way as to render it incapable of being seen in the normal way.

Concluding remarks about hylomorphism, teleology and hypothetical necessity

Metaphysics H4 does not discuss the final cause of sleep, nor does it invoke hypothetical necessity. However, both are discussed in both *Somn.* 2, 455b13–34, and *Somn.* 3, 458a27–32, the two passages that Lowe argued are not part of the main treatise, but rather represent some earlier stage in Aristotle's thinking about the causal explanation of sleep. However, if

[88] *Meteor.* 1.12, 348b2–8; E. W. Webster translation, modified.

[89] The last step in this efficient causal chain is described in the following lines, as part of the definition of sleep, as taking hold of the primary sense-organ (τοῦ πρώτου αἰσθητηρίου κατάληψις at 458a28–29).

the interpretation proposed in the preceding section for the explanation of sleep conforms to the pattern in *Metaphysics* H4, we are now in a position to see that that pattern could be augmented consistently with teleology and hypothetical necessity.

When *Somn.* 2 discusses sleep within the context of natural teleology it does so by invoking his claim that nature acts for the sake of something, and in particular acts for the sake of some good. Aristotle tells us that for things that cannot move by nature continuously and 'with pleasure' it is both good and necessary that they rest. He is not thereby giving an account of the physiological basis for sleep, but rather is appealing to the good of the animal as a teleological cause of its rest. Periods of inability to perceive that are due to sickness, or other deviant causes, would not restore optimal perceptual functioning. To make clear just what type of inability it is that preserves the animal, one would want an efficient causal story that shows how the right kind of incapacity is brought about in such a way as to preserve the life of a perceptually functioning animal, and this is what the explanation I have proposed accomplishes.

It is also something that the definition of sleep at the end of *Somn.* 3 (458a28 ff) purports to have done:

> It has also been stated what sleep is; i.e., that it is a taking hold [κατάληψις] of the primary sense-organ, rendering it incapable of functioning. It occurs of necessity, since it is not possible for an animal to exist, should the conditions that produce it not obtain. Yet it is for the sake of animal preservation. For rest does preserve it.[90]

This is a very particular kind of 'taking hold.' Among other things, it is different from the kind of seizure involved in epilepsy. It is a condition of the primary sense-organ that occurs in regular periods of rest that exactly coincide with the digestion of nutriment. During this part of the nutritive cycle optimal perceptual functioning is no longer possible, and sleep ensues. It lasts just so long as the digestive process is still taking place, and the result of successful digestion is what restores the conditions for the animal's optimal perceptual functioning.

Both by invoking final causality, and by connecting it with hypothetical necessity, these concluding lines of *Somn.* 3 link the regular physiological processes involved both in the digestion of food and the perceptual life of the animal to a broader perspective. Sleep, and its physiological causes, are thereby viewed as part of an ongoing set of processes that regularly and

[90] *Somn.* 3, 458a25–32.

consistently support and preserve the kind of life that is natural for the animal. In order to live, the animal must ingest food, and convert it into blood, the primary form of nutriment for blooded animals. At the phase of the digestive process during which the mixed blood occupies the region surrounding the heart, the heart is temporarily incapable of functioning in its capacity as the primary organ of perception in the way characteristic of waking perceptual life. However, the perceptive faculty is not simply dormant, but instead functions in a different way that is also part of its perceptual functioning. In this condition it is affected by sensory images that have been stored in the sense-organs, and exercises its capacity of imagination both in dreams and other ways. There is no inconsistency between the more limited perspective of *Metaphysics* H4 and this more expansive view of the animal's life. The former is not some later, and more restrictive, methodology that repudiates or replaces teleological explanation and an associated employment of hypothetical necessity. Rather, it can be used to inquire into and to state the physiological causes of sleep in a way that embeds them in a comprehensive teleological perspective.[91]

91 I would like to thank David Ebrey for advice and critical comments that helped bring this out, and to highlight the connection between *Metaphysics* H4 on matter and the physiological account of sleep in the *de Somno*.

CHAPTER 3

Are facts about matter primitive?

Jessica Gelber

In his biological treatises, Aristotle often appeals to some feature of an organism's body in order to explain why it has some other attribute. For example, in *Generation of Animals* II.1 Aristotle appeals to variations in degrees of heat, dryness, and fluidity to explain variations in modes of reproduction – why some animals are livebearers and others are egg-producers. He says that some animals produce "complete" offspring because they are hotter (*thermotera*) and more fluid (*hugrotera*) (732b31–32), whereas other animals produce only "incomplete" eggs since they are either colder (*psuchrotera*) and more fluid, or hotter but drier (*xêrotera*) (733a4–6). In such contexts, Aristotle treats those bodily features as the primary *explanantia* of the relevant *explananda*.

That Aristotle treats certain bodily features – what I will call "facts about matter"[1] – as *explanantia* in particular explanatory contexts is relatively uncontroversial. Recently, however, scholars have been claiming not only that Aristotle uses such facts about matter as *explanantia* in particular explanatory contexts, but also that Aristotle considers those facts to be unexplained.[2] A kind's possession of a certain bodily constitution, for

[1] I have some reservations about calling these facts about *matter*, since I do not think that all of Aristotle's references to elemental forces such as heat and cold are references to the matter of a living organism, if this is understood as that "out of which" its body parts are formed (as, e.g., Ebrey understands it in this volume). I will overlook this for the purposes of this discussion, however. Following those scholars whose claims I am disputing, I will call such features as heat and cold "material" features.

[2] For instance, D. Charles claims that Aristotle "takes as basic" (2000: 334) and treats as "brute physical facts" (*ibid.*: 335) the differences in degrees of heat and fluidity in fish. According to J. Lennox, there are "certain material facts about certain kinds of animals that are as *explanatorily primitive* as are other facts about their living functions" (2001c: 183), and that, in particular, "the amount and kind of elementally distinct materials available in the nutritional make-up of animals is taken as a given" (*ibid.*: 186). A. Gotthelf claims that the fact that lunged organisms have a high degree of heat is, for Aristotle, "explanatorily fundamental" and not "explained in terms of anything more basic about these animals" (1985: 54, n. 24). In the same vein, D. Henry (2008: 59) and M. Leunissen (2010: 97) speak of an organism's material nature making contributions "independently" of the actions of its formal nature. I gather that their thought is that the material nature of a kind not only makes

instance, is said to be "explanatorily primitive" in that (1) the bodily constitution is appealed to in explaining other features of the kind and (2) there is no explanation for why the kind has that particular bodily constitution.[3]

Not every attribute that functions as a starting point in some particular explanatory context is an unexplained or primitive one, of course. For example, we might appeal to the nature of iron to explain why axe blades rust. But that does not entail that it is a primitive fact – one that is not itself explained – that axe blades are made of iron. So, the fact that Aristotle appeals to bodily features (such as temperature or humidity) to explain why a kind of organism has some other attribute (such as its mode of reproduction) does not alone entail that he considers the possession of those bodily features to be primitive. Yet, if Aristotle thinks there is an explanation for the presence of those bodily features, one would hope that he would indicate what that is. And in these specific cases, he does not. Given Aristotle's silence, it is reasonable to conclude that these facts are indeed primitive.

It would be problematic, however, if this were Aristotle's view. First, the idea that facts about matter are "as explanatorily primitive" as facts about soul would constitute a radical departure from views that Aristotle expresses elsewhere. There is some consensus, at any rate, that Aristotle considers explanatory primitiveness – that is, something's explaining other facts but not itself being explained by anything else – to be a guide to essence.[4] However, if an organism's matter were part of its essence, this would be in tension with Aristotle's claim in *de Anima* that soul, not body, is an organism's essence.[5]

Second, the idea that facts about a kind's matter or bodily features are "on the same level"[6] as facts about its form or soul conflicts with the

positive contributions, but that these contributions are completely "independent" of form in the sense that the presence of the matter is not explained or caused by form. If the form explained why that matter is present, matter would not be contributing independently, but rather would be dependent on form.

3 See D. Henry (2013: esp. 238–39 and n. 25) for a clear statement of this view.

4 The connection between explanatory primitiveness and essence is, for example, the subject of the investigation in Charles (2000). As Gotthelf puts the idea, unexplained features are "prime candidates for being part of the essence of the animal in question" (1985: 29). This same assumption is implicit in phrases such as "explanatorily fundamental and thus parts of the *ousia*" (*ibid.*: 53, n. 20) and "explanatorily basic and thus essential" (Gotthelf 1997 [2012]: 86). See also Lennox (2001c: 202): "If Aristotle is inclined to decide what is in the account of a thing's being on the basis of explanatory primitiveness, and if he is willing, in natural science, to include matter in definitions, then we could expect that being blooded or being bloodless would indeed be in the substantial being of animals identified at a sufficiently general level."

5 *DA* II.1, 412b11–13; *DA* II.4, 415b8–12.

6 Charles (2000: 340): "Degrees of heat are taken as explanatory givens on the same level as soul functions, pattern of movement, or being a land animal."

programmatic remarks Aristotle makes within his biological treatises. On the basis of the introductory book of *Parts of Animals*, for example, it is natural to suppose that facts about a kind's matter are going to be explained by facts about the kind's soul:

> Since every instrument is for the sake of something, and each of the parts of the body is for the sake of something, and that for whose sake they are is some activity, it is clear that the composite body, too, is constituted for the sake of some complex activity. For, sawing has not come to be for the sake of the saw, but the saw for the sake of sawing; for sawing is a certain use. So, also, the body is in a way for the sake of the soul, and the parts for the sake of the functions for which each has naturally developed. (*PA* 1.5, 645b14–20)[7]

Rather than indicating that facts about a kind's body or matter are primitive, Aristotle appears to say that facts about a kind's body depend on facts about the kind's soul: The body as a whole, as well as each of the parts of the body, naturally develop for the sake of the psychic functions they are used to perform. So, one expects that in the biological explanations that follow, facts about the form or soul – an organism's psychic capacities and functions – will have explanatory priority over facts about its body. But according to several scholars, this is not what we find.

In response to this, we might simply give up hope of finding consistency between Aristotle's theoretical commitments, as traditionally understood, and his scientific practices. We might conclude that the results of empirical research led to the demise of Aristotle's "brilliant research programme."[8] However, I do not think we need to conclude this. Instead, I here offer three reasons to doubt that Aristotle treats facts about matter – in particular, facts about degrees of heat, dryness, fluidity, etc. – as explanatorily primitive.

First, there is evidence that those putative unexplained facts about degrees of heat, dryness, fluidity, etc. are, in fact, explained. Second, there are certain cases, such as human intelligence, where Aristotle seems to think there is a causal explanation proceeding from facts about matter, but which we have good reason to doubt are intended as proceeding from a primitive fact. Third, the idea that facts about matter are as explanatorily primitive as facts about form requires a particular conception of the causal processes that the explanations mirror. But this conception of the causal

[7] Cf. *PA* 1.1, 642a9–13:

> For, just as it is necessary that an axe be hard, since it needs to cleave, and if hard, made of bronze or iron, so also – since the body is an instrument (for, each of the parts is for the sake of something, and so likewise the whole) – it must be like *this* and made out of *these* materials, if that [for the sake of which they are] is to be.

[8] Charles (2000: 336).

processes conflicts with the way Aristotle characterizes them. In *Generation of Animals*, for example, heat is not treated as a causal factor that works "independently" of soul, but is subservient to soul. In my view, these three considerations provide compelling reason to reject the claim that Aristotle's biology treats facts about matter as primitive.

Hot, cold, earthy, and fluid natures

When Aristotle talks about a type of organism's matter, he is sometimes talking about its body, in contrast to its soul. Other times, he describes the parts of which the whole body is composed as the matter. Those body parts, moreover, are discussed at three levels of generality. There are the "non-uniform" parts, such as limbs and organs, the "uniform" parts, such as flesh and bone that the limbs and organs are made out of, and the elemental potentials that make up the latter. For instance, at the beginning of *GA* Aristotle puts the point as follows:

> . . . and the matter in animals is the parts, the non-uniform parts [are the matter] for every whole [animal], the uniform parts [are the matter] for the non-uniform, and the so-called elements of the bodies [are the matter] for these. (*GA* I.1, 715a9–11)[9]

Not only the body as a whole, but also the non-uniform limbs and organs, the tissues, bone, and other uniform parts those are made of, and the elemental bodies (or, more precisely, the elemental powers) are considered by Aristotle to be an organism's matter. Here I will focus on matter at that lowest level, the bodily blend or *krasis* of elemental powers in an organism's blood (or the analogous substance in non-blooded organisms). I will focus on matter at this level because it is the strongest candidate for being a primitive fact about matter. For, as the lowest level of composition, the elemental powers are the matter out of which the primary uniform part, blood (or the analogous substance in bloodless animals), is composed. Since blood is constituted by a particular combination of elemental powers, differences in blood are due to differences in the proportions of the elemental powers composing it, that is, by its *krasis*. For instance, blood can be more fluid

[9] Cf. *PA* II.1, 646a12–24 (Lennox trans.):

> Since there are three compositions, one might put first composition from what some people call the elements, e.g., earth, air, water, and fire. And yet, perhaps it is better to speak of composition from the potentials . . . That is, moist, dry, hot, and cold are the matter of the composite bodies . . . Second is the composition of the nature of the uniform parts within the animal – e.g., of bone, flesh, and the other things of this sort – out of the primary things. Third and last in the series is the composition of the nature of the non-uniform parts – e.g., of face, hand, and such parts.

or more solid, hotter or less hot, depending on its proportion of elemental powers. Blood, moreover, is the matter and nourishment for the rest of the parts,[10] and thus blood (or the analogous uniform part in bloodless organisms) is said to be the cause of many other features.

> The nature of the blood is the cause of many features of animals with respect to both character and perception, as is reasonable, since blood is the matter of the entire body; for nourishment is matter, and blood is the last stage of nourishment. It therefore makes a great difference whether it is hot or cold, thin or thick, turbid or pure. (*PA* II.4, 651a12–17)

In Aristotle's view, characteristics such as the "heaviness, lightness, density, rarity, roughness, and smoothness" of parts and limbs (*PA* II.1, 646a18–20) "follow from" the particular *krasis* that makes up the blood and so determines its quality. More earthy blood will give rise to parts that are, for instance, rougher rather than smooth.[11] In addition, *krasis* of blood is correlated with character traits (*ethê*). For example, Aristotle says that timid animals have excessively watery blood, whereas organisms with more earthy blood are more passionate (*PA* II.4, 650b27 ff). Further, there is some indication that Aristotle thinks *krasis* determines what kind of food an organism eats, or at least what it finds pleasant (*HA* VII[VIII].2, 589a8–9; 590a10–11).[12]

A kind's *krasis*, then, is identified by Aristotle as a cause of many other attributes of the kind. Characteristics such as skin texture, aspects of the kind's character, and phenomena constitutive of the kind's life (such as its mode of reproduction) all appear to be due to the particular *krasis* of blood that organisms of that kind have. Is a kind's possession of a particular *krasis* of blood a primitive, unexplained fact about the kind?

If one accepts the following principle, the answer is no:

> **P**: If X is said to be present for the sake of Y, then X's presence is explained by Y.

Aristotle's views about explanation are complex, but I think it is safe to assume that he thinks that one explains why something is the case by citing causes. And, famously, Aristotle recognizes "that for the sake of which" as a type of cause. Given this, it follows that to cite the purpose for which

[10] *PA* II.3, 650b2–3, b12–13: "It is, then, apparent from these and like considerations that blood is present in blooded animals for the sake of nourishment . . . blood is for the sake of nourishment, i.e. the nourishment of the parts."

[11] See *PA* II.9, 655a26–28 with IV.13, 697a7–9 on the selachian's rough skin.

[12] See also the various remarks about the connection between *krasis* and an organism's way of life in VII[VIII].2.

something is the case is, for Aristotle, to render that thing at least partially explained.[13]

Assuming that Aristotle is committed to P, then the discussion of the levels of composition in *PA* II.1 is evidence that a kind's *krasis* of blood is not, for Aristotle, explanatorily primitive. For, Aristotle explicitly says there that a kind's *krasis* is for the sake of something.

> Thus the matter of the elements is necessary for the sake of the uniform parts, since these are later in generation than the elements, and later than the uniform parts are the non-uniform parts. (*PA* II.1, 646b5–8 (Lennox trans.))

Let me state very briefly what I take Aristotle to be saying in the surrounding context. The highest of the three levels of material composition of living bodies – the instrumental, non-uniform parts such as eyes and hands – have functions. And it is for the sake of performing those functions that they are present in an organism. For instance, birds have wings because they are flyers, since wings are the instruments required for flying (*PA* IV.12, 693b11–13). And sensory organs are present because they are required for performing perceptual activities. Flesh, for instance, is the organ of the primary sense – touch – that animals must have since they are by definition *perceptive* living things (*PA* II.8, 653b22–24).[14]

Most variations,[15] such as variations in size or quality of the instrumental parts and sensory organs, are explained by the functional needs of the particular kind of organism.[16] For instance, some birds (such as the carnivores

13 I say "at least partially" because Aristotle says at *APo.* II.11, 94a20–27 that one should give demonstrations "through" each of the four causes to fully explain something. I cannot address here the complicated issues about Aristotle's conception of explanation. It is sufficient for my purposes to note that there is a difference between cases in which there is a demonstration through some cause or other, and cases in which there simply is no cause to cite and so no demonstration to give, for example, in the case of indemonstrable statements of essence. There will be material and efficient causal explanations of *how* organisms perform the activities constitutive of their essence, of course, but that does not render the fact *that* the organisms have those essential activities explained. That is, there is no explanation of why they have the essential activities and functions that they do.

14 It is uncertain how, exactly, Aristotle is thinking of the role that flesh plays in tactile sensation. See Lennox (2001a: 213) for a discussion of this. At any rate, it is clear that Aristotle thinks that flesh is necessary for touch, so this complication does not matter for my purposes here.

15 Two exceptions come to mind: Aristotle says that a certain octopus has a slender body and that the elephant has a large size, but does not give any explanation for these facts. The slenderness of the octopus explains its having only a single row of suckers (*PA* IV.9, 685b12–16), and the elephant's large size explains its having a trunk (*PA* II.16, 658b33 ff). It is not clear how Aristotle is conceiving of size or other "dimensional" features, though there is some reason to be cautious about calling these features "material" ones. In *de Anima* II.4, 416a16–18 Aristotle claims that the "limit and *logos* of size and growth" is "of soul" and more "of account" than "of matter."

16 In other cases, differences are explained by the kind's *krasis*. The number of gills that a kind of fish has, for instance, is due to its degree of heat (since gills are the instruments for cooling in

with crooked talons) have larger wings with plentiful feathers (*PA* IV.12, 693b27–694a4) because they need to be good flyers (presumably because of the way they catch their food, by chasing down rodents and grabbing them). In order to perform their functions, the instrumental parts must have certain powers or capabilities. The human hand, for instance, "needs one power (*dunamis*) for the action of compressing and another for that of grasping" (*PA* II.1, 646b23–25). In order to have such powers, the hand must be made of appropriate uniform matter, for example, matter that has what I will call "dispositional properties" such as hardness, softness, elasticity, and flexibility.[17] Uniform parts, considered as parts with certain dispositional properties, are the parts that meet the functional requirements of the non-uniform parts they compose.[18] This is why the uniform parts are said to be for the sake of the non-uniform.

Finally, the dispositional properties of uniform parts are brought about by particular kinds of elemental blends. As we learn in *Meteorology* IV, dispositional properties (called *pathêmata* in *Meteorology*) are the results of the interactions of the active powers – the hot and the cold – and the passive powers – the moist and the dry. Particular proportions of those elemental powers give rise to those dispositional properties by which uniform substances such as flesh, bone, and wood are distinguished from one another, such as solubility, solidity, and flexibility. When a particular elemental blend is the right one to bring about the uniform parts (considered

water-dwellers, and so hotter ones need more gills to cool them) (*PA* IV.13, 696b12–23). But, again, the fact that *krasis* explains other features does not entail that the presence of that particular *krasis* is a basic, unexplained feature.

[17] *PA* II.1, 646b10–27:

> Now animals are composed out of both of these two sorts of parts, uniform and non-uniform; the former, however, are for the sake of the latter, as it is to the latter that actions and operations belong (e.g. eye, nose, the face as a whole, finger, hand, the arm as a whole). And inasmuch as the actions and movements both of an animal as a whole and of its parts are complex, the substances out of which these are composed must of necessity possess distinct *dunameis*. Softness is useful for some purposes, hardness for others; some parts must be able to stretch, some to bend. In the uniform parts, then, such *dunameis* are found apportioned out separately: one of the parts, for instance, will be soft, another hard, while one is fluid, another solid; one viscous, another brittle. In the non-uniform parts, on the other hand, these *dunameis* are found in combination, not singly. For example, the hand needs one *dunamis* for the action of compressing and another for that of grasping. Hence it is that the instrumental parts of the body are composed of bones, sinews, flesh, and the rest of them, and not the other way around.

[18] Lips and nostrils, for instance, are "fleshy" (*sarkina*) because the dispositional features of flesh are required by the functions that lips and nostrils perform in living organisms that have them (see *PA* II.16, 660a8–11). The function of lips in all animals that have them is to protect the teeth. In humans, in addition to this function, lips also help enable speech (*PA* II.16, 659b30–34). The function of nostrils in organisms that have lungs is to enable breathing (*PA* II.16, 659a30–31).

as parts with certain dispositional features), that blend is required by, and so present for the sake of, the presence of the uniform parts.

Each level of bodily composition is said to be for the sake of the one above. The matter for the instrumental parts – the uniform parts – must have specific dispositional features appropriate to their functions. And the uniform parts, considered as parts with distinctive dispositional features, must be composed of specific elemental blends that give rise to those dispositional features. For this reason, the particular elemental composition of the blood is for the sake of the parts that come to be formed out it. Thus, if we accept principle P, Aristotle's discussion in *PA* II.1 gives us grounds for thinking that a kind's *krasis* is not primitive.

In sum, a kind's particular *krasis* can give rise to various features (such as hard eyelids) because of the nature of that elemental blend, and thus *krasis* can explain (in a given explanatory context) other features. And certain features, such as the dispositional properties that the uniform parts need to have, may come to be formed "of necessity" (as described in *Meteorology* IV), because the material at the lowest level of composition naturally has certain effects. But this does not make it a "brute" or "primitive" fact that an organismal kind has that particular composition or blend. It might be a brute or primitive fact that the elements have the causal powers that they do, but asking why the elements have the causal powers they do is different than asking why some particular elemental blends are present in a kind of organism's blood.[19] As I understand it, Aristotle's view is that the presence of a particular *krasis* in an organism's blood is explained teleologically, by reference to the dispositional properties that the kind's uniform parts must have. Uniform parts must have certain dispositional properties, given the functions that the non-uniform parts they constitute must perform. If this is correct, then a kind's *krasis* is not a primitive, unexplained fact.

Heat and human intelligence

The argument offered in the last section relied on the claim that *krasis* is said to be for the sake of something in *PA* II. According to what I called principle P, this renders the presence of the *krasis* explained, and not primitive. But one might deny that Aristotle is really saying that *krasis* is for the sake of something in the strong sense that makes P sound like a

[19] On this point, see Cooper (1987: 261): "if an animal of a certain kind is to be constituted these certain amounts of certain elements must be present for use: in effect, for *this* creature *those* elements are hypothetically necessitated. But plainly that presupposes, and does nothing to explain, the natural powers of the elements concerned."

plausible principle. In this section I offer a second reason for rejecting the idea that the possession of a particular *krasis* is primitive, which reason does not rely on the claim that *krasis* is for the sake of the parts formed out of it.

Suppose that Aristotle thinks that a kind's *krasis* explains why the kind has the mode of reproduction that it does, and that there is no further explanation for the kind having that particular *krasis*. Suppose the possession of a particular *krasis*, that is, is explanatorily primitive. Now explanatory primitiveness in the sense at issue here is not relative to an explanatory context. So, if the particular *krasis* an organism has is being treated as a primitive fact about the kind in the explanation of its mode of reproduction, then it must also be primitive in any other explanation. Yet, some of the causal connections that Aristotle cites between a kind's *krasis* and other features are such that it would be surprising if *krasis* – as opposed to the feature that the particular *krasis* is causally connected to – were being treated as explanatorily primitive. Consider, for example, the connection between *krasis* and human rational activity:

> . . . in man the brain is more fluid and greater in volume than in any other animal, and the reason of this, in its turn, is that the heat in the heart is purest in man. The fineness of the blend (*eukrasian*) in man is shown by his possession of intellect: there is no other animal which is so intelligent. (*GA* II.6, 744a26–31 (Peck trans.))

Humans, Aristotle says, have the highest degree of heat of any living organism. That we have the most heat explains, among other things, why humans alone have an upright posture.

> For heat promotes growth, and a profusion of blood is a sign of heat. And further, the bodies of those that are hotter are more erect, which is why mankind is the most erect of the animals. (*PA* III.6, 669b3–6 (Lennox trans.))

Being able to stand upright, moreover, allows humans to engage in rational activities:

> Mankind, however, instead of forelimbs and forefeet has arms and what are called hands. For it alone of the animals is upright, on account of the fact that its nature and substantial being are divine; and it is a function of that which is most divine to reason (*noein*) and to think (*phronein*). But this is not easy when much of the body is pressing down from above, since the weight makes thought and common sense sluggish. (*PA* IV.10, 686a25–32 (Lennox trans.))

And engaging in rational activities is what distinguishes humans from other animals. Being able to do so is part of human nature, and actually doing so is our natural end or purpose (*telos*):

> Reason (*logos*) and intellect (*nous*) are for us the goal of nature. (*Pol* VII.15, 1334b15)[20]

It is reasonable to suppose that a kind's *telos*, for example that the human *telos* is rational activity, is a primitive fact. For, what something is, essentially, and what it is for the sake of, Aristotle says, are the same.[21] And what something is, essentially, has no further explanation.[22] If possessing a high degree of heat were simply a primitive fact about humans, rather than something that is conditionally necessary for rational activity, then part of the essence of human beings (rational activity) and a fact about the human body's material composition (high degree of heat) would be two independent facts. That the two facts are coordinated so well – that is, that the material composition is conducive to some part of the essence – would then turn out to be due to chance. But this fortuitous coordination occurs regularly, and Aristotle denies that regularly occurring, beneficial outcomes could be merely due to chance.[23]

Further, it is not solely human rationality that Aristotle correlates with the *krasis* of a kind's blood (or the analogous substance in bloodless organism). First, differences in *krasis* are also linked to the intelligence and perceptive capacities of non-human organisms, such as bees:

> Thicker (*pachuteron*) and hotter blood is more productive of strength, while thinner (*leptoteron*) and cooler blood is more perceptive and intelligent (*noerôteron*).[24] And the same difference obtains among the attributes analogous to blood. This is why both bees and other such animals are wiser (*phronimôtera*) in their nature than many blooded animals, and why among blooded animals those having cold and thin blood are wiser than their opposites. But those with hot, thin, and pure (*katharon*) blood are best; for such animals are at once in a good state relative to both courage and wisdom (*phronêsin*). (*PA* II.2, 648a2–11 (Lennox trans., modified))

20 Cf. *NE* X.8, 1178b21–23: "So, then, the activity of a god, superior as it is in blessedness, will be theoretical activity; and so, too, the human activity that has the greatest affinity to this one will be more productive of *eudaimonia*."

21 *Phys.* II.7, 198a24–27. Cf. *GA* I.1, 715a8–9: "the *logos* and that for the sake of which as end are the same (*tauton*)."

22 Of course, *how* a kind achieves its *telos* will have a material and efficient causal explanation. But *what* it is does not.

23 Aristotle argues in *Phys.* II.8 against an opponent of natural teleology by arguing that regularly occurring, beneficial natural phenomena cannot be due to chance, and that consequently they are finally caused.

24 I am taking references to "thick" and "thin" blood as references to blood that is drier and more fluid, respectively. In *GC* II.2, 329b17–330a10, Aristotle says that thick (*pachu*) and thin (*lepton*) are 'of' dry and fluid, respectively. What exactly this amounts to is not important for my purposes. I am solely interested in the connections drawn there between thickness and dryness, and thinness and fluidity.

And heat seems to be correlated with other ways that Aristotle ranks animals:

> The reason why some creatures have this part, and why those having it need respiration, is that the more honorable animals have a greater proportion of heat. (*de Resp.* 477a14–16 (Ross trans.))

In my view, Aristotle is not ordering animals on the scale of honor or value on the basis of how much heat their bodies happen to have, as a primitive fact. If this is so, we have good reason to doubt that Aristotle considers a kind's *krasis* to be primitive.

Tools

In this section, I offer a third reason for doubting that Aristotle considers facts about matter to be primitive. In short, this reason is that in Aristotle's biological account of animal generation, the elemental powers such as heat and cold do not operate independently of, but rather are subordinate to, an organism's soul. Given that elemental powers play this subordinate causal role, I will argue, facts about matter cannot be primitive in Aristotle's biological explanations.

Before arguing for this, I should emphasize that the idea that Aristotle thinks facts about matter are primitive is by no means unmotivated. For example, Aristotle claims in *Physics* II.2 and *PA* I.1 that it is incumbent on the natural scientist to understand the matter as well as form, since both the matter and form are said to be a thing's nature. A natural scientist, Aristotle thinks, must understand the material basis for the exercise of the various vital capacities (with the exception of *nous*) that make up an organism's soul and essence. One cannot make sense of life and living things without understanding the nature "as matter." Aristotle also claims that natural beings are analogous to "the snub" in that they cannot be defined without reference to the matter, just as snubness cannot be defined without reference to the nose. And there are indications that Aristotle considers some bodily features to be in the being (*ousia*) or included in the definition (*logos tês ousias*) of the kind.[25] When we add to all of this Aristotle's appeals to the necessary interactions of heat and cold, for instance, in explaining various features of living organisms, it is not unreasonable to think that Aristotle is treating facts about matter as explanatorily primitive.

[25] Gotthelf (1985) discusses in detail the passages where some bodily feature seems to be treated as included in the essence or definition. See, however, note 15, above.

The idea that matter is explanatorily primitive, however, requires a certain picture of the causal processes that the scientific explanations mirror. On the interpretation of material facts as primitive, there are, as it were, two independent sets of explanations; some explanations proceed from form or soul, and some proceed from matter. Corresponding to those explanations that proceed from the form or formal nature there is one causal process, and corresponding to those explanations that proceed from matter is another. Proponents of the interpretation of material facts as primitive think that in addition to the goal-directed causal processes, there is also "a necessity rooted in the material nature of an animal, which constrains, and perhaps acts independently of, the actions of its formal nature."[26]

There is an alternative picture, however, that more accurately portrays matter's role. This alternative can accommodate the fact that the matter of a living organism is in a sense its nature, insofar as it is a source of change and rest that belongs to it non-accidentally. And this picture also allows for certain features of living organisms to be formed "of necessity" because of the interactions of the material elements. Nevertheless, this alternative does not construe matter as operating independently of form or soul, or treat the presence of matter as primitive. This is the picture that emerges in *Generation of Animals*.

In *GA*, Aristotle is explaining how the efficient cause of generation (the father or the father's nature or soul) conveys the form (which he has in actuality) to the matter provided by the mother, and how the embryo is then formed. To do so, Aristotle must explain the material basis by which form is conveyed, and then how the embryo's tissues, bones, limbs, and organs are constructed. That material basis, as it turns out, is what we might think of as metabolic processes – heating and cooling – in the spermatic residues and nutritive (and growth promoting) blood.[27]

When Aristotle describes this process, he sometimes refers to heat and cold as the *tools* of soul. Elsewhere, Aristotle also describes the body as a tool of soul.[28] Insofar as Aristotle is conceiving of soul as a living substance's capacities for engaging in vital activities, it is natural to understand such descriptions as indicating that the body parts are the means or instruments by which vital activities are performed, much like an artisan's tools are the means by which the artisan's *technê* is carried out. And for parts that have functions, the description seems apt. Eyes, for instance, are the tools by

26 Lennox (2001c: 187).

27 See, for example, *GA* II.6, 743a36–b5 and *GA* II.1, 734b28–735a4, both quoted below.

28 For example, *PA* I.1, 642a9–13; *PA* II.7, 652b7–15; and *DA* 2.4, 415b18–20: "all natural bodies are tools of soul, as those of animals, so too, those of plants, as being for the sake of soul."

which an organism engages in seeing; hands are tools by which we grasp and press things. But heat and cold are less obviously analogous to the craftsman's tools than are functional organs and limbs. And Aristotle is aware of this, which is why he modifies the analogy with craft production when explaining how nutritive soul uses heat and cold as tools:

> . . . as the products of art are made by means of the tools of the artisan, or to put it more truly by means of their movement, and this is the activity (*energeia*) of the art, and the art is the form of what is made in something else, so is it with the power of nutritive soul. As later on in the case of mature animals and plants this soul causes growth from the nutriment, using heat and cold as its tools (for the movement of the soul is in these), and each thing comes into being in accordance with a certain formula, so also from the beginning does it form the product of nature. (*GA* II.4, 740b25–34 (Platt trans., modified))

Here Aristotle compares heat and cold not to an artisan's tools but to their *movements*, that is, the tools as they are being used. These movements of the artisan's tools, Aristotle says, are the *energeia* or activity of the artisan's *technê*. This suggests that the important point of the analogy between the movements of the artisan's tools and the heating and cooling in blood is that each is the activity or *energeia* of some *dunamis*. In living organisms, the heating and cooling are the activities of soul capacities; these are how those vital activities are carried out.

If we grant that Aristotle is thinking of heat and cold in this way – that is, as tools by which soul capacities are exercised – two points follow. First, it follows from the idea that heat and cold are tools or instruments that their causal role is thereby a subsidiary one. They are *sunaitia*, not *aitia*.[29] And subsidiary causes never operate independently of the first or primary cause to which they are subordinate. Consequently, as subsidiary causes, the elemental powers in living organisms' bodies do not operate independently of, but are rather subordinate to form or soul.[30]

Second, auxiliary causes can have their own *per se* effects. The drugs that the doctor prescribes, for instance, have powers "of necessity" to alter the patient's body. Those powers the drugs have are not working independently of the doctor's art, however; the drugs are present in the patient's body (they have been ingested, that is) only because the doctor prescribed them. Nevertheless, the powers in the drugs cause changes in the patient.

[29] Cf. *DA* II.4, 416a13–14.

[30] Balme (1987: 276) also makes the point that as *sunaitia*, the matter's "natural action and movement" will not produce its effects unless "nature causes it to do so."

Similarly, heat and cold in living organisms have powers, grounded in necessity, of producing features (softness, hardness, etc.) of the body, just as they do in craft production.

> Heat and cold (which is deprivation of heat) are both employed by Nature. Each has the power, grounded in necessity, of making one thing into this and another thing into that; but in the case of the forming of the embryo it is for a purpose that their power of heating and cooling make it such, partly owing to necessity, partly for a purpose – sinew solid and elastic, bone solid and brittle. (*GA* II.6, 743a36–b5 (Peck trans.))

As tools of soul, and so subsidiary causes, heat and cold can cause changes in the embryo's body. But still, just as in the case of craft production, heat and cold do not operate on their own.

> And as in speaking of an axe or any other instrument, we should not say that it was made solely by fire, so we should not say this about a foot or a hand, nor, similarly, of flesh either, because there is a function of this also. As for hardness, softness, toughness, brittleness and the rest of such qualities which belong to the parts that have soul in them, heat and cold may very well produce these, but they certainly do not produce the *logos* in virtue of which the one is now flesh and the other bone. Rather, the movement derived from the generator who is in actuality that which the material out of which the offspring is formed is in potential. The very same thing applies to things formed in accordance with art. For, heat and cold may soften and harden the iron, but they do not produce the sword. This is done by the movement of the tools, which has the *logos* of the art. For the art is the principle and form of the thing being made, but in another. But nature's movement is in [the product being formed], derived from another natural being having the form in actuality. (*GA* II.1, 734b28–735a4)

Viewed from this perspective, the idea that matter operates independently of soul does not really get a foothold. Rather, matter is always operating as a tool of soul, and so subordinate to it. Accordingly, facts about matter in the scientific demonstrations that display the causes will also be subordinate to facts about form or soul, and not "as explanatorily primitive" as them.

Conclusion

I began by noting a recent trend to interpret Aristotle's biological explanations as treating facts about matter as explanatorily primitive. This interpretation, however, is in tension with what we would have expected, given what Aristotle says outside the biological works. And, I argued, it is far from obvious that the biological explanations treat facts about matter as primitive. I offered the following three considerations.

First, from Aristotle's discussions in *PA* II.1, it would seem that matter at the lowest level of composition – an organism's bodily blend or *krasis* – is not primitive, at least not if one accepts that citing what something is for the sake of entails that it is explained. Second, there is reason to think that certain appeals to *krasis* are not intended as references to primitive facts. For Aristotle correlates facts about a kind's *krasis* with facts about human rationality. If human rational capacities are part of human essence, this would render the coordination between essence and matter a coincidence. But I do not think that this can be Aristotle's view. Third and finally, I have tried to indicate the way in which Aristotle's conception of the metabolic processes – the heating and cooling involved in reproduction, development, and maintenance – imbeds the actions of the elemental powers into the activity of the soul or form of the organism. Heat and cold are tools by which soul activities are carried out. This bars them from being "independently" operative in the way that has been suggested in the recent literature.

CHAPTER 4

Blood, matter, and necessity

David Ebrey

According to most scholars, in the *Parts of Animals* Aristotle frequently provides explanations in terms of material necessity, as well as explanations in terms of that for the sake of which, that is, final causes.[1] This material necessity is not supposed to be the necessity from a hypothesis that Aristotle describes in *Physics* II.8, which involves the necessity of matter given some end, but rather a type of necessity that results directly from matter, for example, the necessity of fire to heat. Aristotle says in the first chapter of the *Parts of Animals* that he is looking for two causes, the "of necessity" and the "that for the sake of which" (642a1–3), and he follows through on this in the rest of the *Parts of Animals*, frequently bringing his explanations back to these two causes. But he never uses the phrase "material necessity." It is a substantive interpretation that matter is connected to necessity in this way.

In this chapter, I argue that we misunderstand both matter and the way that Aristotle explains things using necessity if we interpret Aristotle as explaining things in terms of material necessity.[2] In the first half of the chapter, I provide an account of how Aristotle uses the term "matter" in the *Parts of Animals*. He does not use it very frequently in his detailed discussions of animal parts; when he does use it, he typically identifies blood as matter. The reason for this, I argue, is the following. Blood is, for Aristotle, what properly nourishes and grows the other parts of the body and he views nourishment and growth as types of coming to be. Matter, for Aristotle, is what comes-to-be other things and hence what nourishes and

I have received valuable feedback from Devin Henry, Northwestern's Junior Faculty Working group, UW-Milwaukee, The Greeks at UW-Madison, the conference "Theory and Practice in Aristotle's Natural Philosophy" at Northwestern, and at the "New Directions" workshop at Rutgers.

1 Lennox 1987, Gill 1997, Henry 2007, and Leunissen 2010 (*passim*). This is separate from the view Gelber discusses in this volume that there are primitive facts about matter. However, these views are typically found together and naturally complement one another: there are primitive facts about matter and these are necessary facts.

2 Lennox's (1987), Henry's (2007), and Leunissen's (2010) projects are not primarily focused on matter or necessity. So, even if I am right, this need not cause major problems for their overall projects.

grows the parts of the body will be a type of matter. Blood, for Aristotle, can nourish or grow different animal parts and so, rather than necessitating a particular outcome, it is open to very different results.

The second half of the chapter turns to necessity as a cause in the *Parts of Animals*. How should we understand this, if we do not think of it as having a special connection to matter? I argue that sometimes this necessity is the result of agent and patient coming together. But in other cases the necessity follows from a thing's essence or is a sort of conceptual truth. Aristotle does not distinguish between these different sorts of necessity in the *Parts of Animals*. In fact, he shows no interest in analyzing the types of necessity involved in his detailed biological explanations. Instead, he views his task as complete once he has brought a question back to a necessity as a cause (perhaps along with a final cause). The term "material necessity" obscures the unusual sort of cause he is giving in the *Parts of Animals*: undifferentiated necessity.

Blood and matter in the *Parts of Animals*

It is important not to allow our ordinary ideas about matter to influence how we understand matter in Aristotle. The term we translate as "matter," *hulē*, originally meant trees, wood, or lumber. The only extant use before Aristotle to mean something roughly like matter is a single occurrence in Plato's *Philebus* (54c). Aristotle certainly developed his notion of *hulē* using related ideas from his predecessors, and perhaps other people were developing ideas about *hulē* in the Academy. Nonetheless, Aristotle was at the forefront of developing a new concept, one that he says in *Metaphysics* A was not properly grasped by any of his predecessors (985a9–17 and 993a11–17). It is not clear how close our current ideas about matter are to the one that Aristotle thought was crucial for understanding the natural world.

Given this, it is useful to work our way up from the passages where Aristotle uses the term *hulē*. Aristotle only uses the word 30 times in the *Parts of Animals*; 15 of them are in the first, theoretical, book; 5 of them are in the first page of book II, chapter I, leaving only 10 occurrences in the rest of books II–IV, where Aristotle provides his detailed discussion of animal parts.[3] This is compared to 68 occurrences in the *Generation of Animals*. Moreover, in 7 of the last 10 occurrences, the matter in question is identified as blood.

[3] Based on searches in the TLG.

Why does Aristotle generally identify blood as matter in his detailed discussion of animal parts? While there are complications, the basic reason is this: matter, for Aristotle, is intimately tied with change. In particular, the matter for some change is what properly speaking undergoes that change. I argue elsewhere that this is the correct interpretation of matter in Aristotle's *Physics* book I.[4] In brief, Aristotle motivates the need for matter in *Physics* I by arguing that in every change there must be something that properly undergoes the change. This is why he says in *Phys.* I.7 that things come to be from matter not by virtue of concurrence (190b24–27). In this chapter I simply argue that what we find in the *Parts of Animals* fits with this view. Blood, according to Aristotle, is produced by the body precisely to be something that can come-to-be each of the internal organs and uniform parts. Matter plays a more prominent role in the *Generation of Animals* because generation is central to that work.

Blood is the first part that Aristotle discusses in the *Parts of Animals*.[5] He begins his discussion of blood in *PA* II.2, but says that to discuss it he first must discuss heat and the opposites dry and moist, since blood is hot and moist. He concludes this preliminary discussion, in II.3, with the claim that blood is the final nourishment in blooded animals (bloodless animals, he thinks, have an analogue to blood). There is the nourishment that we take into our bodies as food. But this must be concocted, with heat, in the stomach, until it is in the form of this final nourishment. Blood, he says, is for the sake of nourishing. Blood's moisture and heat are not independent features of it; it nourishes because it is moist and hot.[6]

Aristotle thinks that some of an organism's differing characteristics are explained by different types of blood: some animals have a subtle intelligence because of thin and pure blood; timid animals have watery blood; those with fibrous thick blood are more spirited (648a2–11; 650b19–651a5). Near the end of discussing this feature of blood he connects it to matter:

> The nature of the blood is the cause of many features of animals with respect to both character and perception, as is reasonable, since blood is the matter of the entire body; for nourishment is matter and blood is the last stage of nourishment. (651a12–15)[7]

4 See Ebrey 2007 and Ebrey (unpublished) "Distinguishing Matter from Lack in *Physics* I." See also Alan Code's contribution in this volume, which discusses Aristotle's view of matter in *Metaphysics* H and is complementary to my own.

5 Chris Frey (forthcoming), "From Blood to Flesh: Homonymy, Unity, and Ways of Being in Aristotle," *Ancient Philosophy*, section I, argues that Aristotle's considered view is that blood is not a part. For my purposes, it does not matter whether it is a proper part or merely a part-like thing.

6 For a short, complementary discussion of blood in Aristotle, see Freeland 1987, p. 401.

7 Translations from Lennox 2001a, occasionally lightly revised.

Aristotle says that blood is the matter for the entire body because it is nourishment for the body. In order to explain why blood's being nourishment qualifies it as matter, we need to understand Aristotle's theory of nourishment. In the *Parts of Animals* he refers us to discussions of nourishment (650b10, 668a8–9). This may refer, in part, to a lost treatise; however, there is a discussion in *Generation and Corruption*'s chapter on growth (I.5) and a more extensive discussion of nourishment throughout book II of *Generation of Animals* (especially II.4, II.6, and II.7).[8] Nourishment is treated as fundamentally similar to growth (e.g., *GC* I.5, 322a23–28). In both cases, the body takes something into it. Normally, if something enters a body, it is simply a foreign body. But in growth or nourishment, the thing that is brought in becomes part of the new thing. Thus, growth and nourishment are both types of coming to be. The difference is that in growth there is a change of size, whereas in nourishment the thing does not change in size (typically, one imagines, because waste is also produced).

Blood is concocted to be the body's nourishment. But, in fact, it is more than that. The organs themselves come to be from blood, as we can see in the following passage from *PA* III.5, Aristotle's chapter on blood vessels:

> A cause of the blood vessels being distributed over the entire body is that the blood and its analogue in bloodless animals are matter for the entire body, while these are stored in blood vessel and its analogue. As to how and from what animals are nourished, and in what manner they absorb nourishment from the gut, it is more appropriate to investigate and speak about such things in the works on generation. And since the parts are constituted from the blood, as we said, it is reasonable that the course of the blood vessels runs naturally through the entire body; for the blood too needs to be passing through everything and next to everything, if each of the parts is to be constituted from it. (668a1–11)

At the beginning of the passage we see that the fact that blood is matter for the entire body explains why we have blood vessels throughout our entire body. And at the end of the passage we are told that each of the parts is constituted by blood.[9] Thus, blood constitutes, nourishes, and grows the parts of the body. In *GA* IV.1 Aristotle ties this to a more general principle:

[8] For a discussion, see Lennox 2001a, pp. 200–1.
[9] Note that Aristotle thinks that different types of blood are involved in different types of change.

> Since it is the same, that out of which something comes to be and grows, namely the nourishment, each of the parts comes to be out of that sort of matter and that sort of residue which it is fitted to receive. (766a10–13)

Note, then, that blood is quite open to possibility for Aristotle – blood can come to be, nourish or grow a number of different parts; on its own, blood does not necessitate any particular action.

We have seen that in the detailed discussion of parts in the *Parts of Animals* Aristotle typically identifies matter as blood. This is what the body concocts to properly nourish and grow the uniform parts. In general, things properly come to be from matter and Aristotle thinks that the body only concocts one thing to play that role: blood. However, there are a few cases in the detailed discussion in the *Parts of Animals* where something other than blood is identified as matter. There is the case of birds lacking ear-matter considered in the next section of this chapter. For now, let us consider passages from the beginning of book II that suggest a progressively broader scope for matter. Near the beginning of Aristotle's discussion of blood he suggests a somewhat broader scope for matter:

> First of all, then, many modes of cause may be attributed to the moist parts and the dry. Some of them serve as matter for the non-uniform parts (since each of the instrumental parts has been constituted from these, i.e., from bones, sinews, flesh, and other such parts, some contributing to the *ousia* [substance, being] *of* the instrumental parts, some to their operation); some of the moist parts are nourishment for the non-uniform parts (for all derive their growth from what is moist); and some of them turn out to be residues, such as the sediment from dry nourishment and, in those with a bladder, from moist. (647b20–29)

Bones, sinew, flesh, and other parts are described as matter, and put alongside the moist parts that are nourishment. Since blood is a moist part that is nourishment, this might even suggest that blood is not matter, although Aristotle does not actually say this (and it contradicts what he says later in the chapter).[10] Aristotle's account seems to be the following: blood is the matter for the uniform parts and nourishes all parts, whereas the other uniform parts (such as bones, etc.) are, in turn, matter for arms, legs, etc. When blood nourishes the arm, it does so by nourishing the uniform parts, such as flesh, that constitute the arm.

[10] At the beginning of *GA* I.1 all parts are described as matter.

This leaves us with a question: If bones and sinews are matter for the non-uniform parts, why does Aristotle not mention this in his discussion of bones and sinews? Why is blood repeatedly referred to as matter in the *Parts of Animals*, and not these uniform parts, if they are also matter? The reason is that there is something special about blood: its nature is to come to be something else. The other parts are not as intimately tied to coming to be as blood. While these parts do play a role in the coming to be of the whole organism, their primary function is not to play this role. Their primary function is determined by their function in a fully formed organism. One can provide an account of bone without mentioning that it is matter, in a way that one cannot do with blood. The primary function of blood is to generate, maintain, and grow the parts, and these are all types of coming to be. For blood to be fully developed is for it to be primed to come to be something else, unlike with the other parts.

On this account, something is matter to the extent that it is playing a certain role.[11] When Aristotle is describing something, but not as part of this role, he is not describing it insofar as it is matter. Aristotle would not tack onto his description of the liver that it is matter for the organism. This is not relevant to the sort of part-by-part description he gives in the *Parts of Animals*. But when Aristotle asks whether he has discussed the matter of the organism (as he does at the beginning of *GA* I.I), he can say that he has. This is one important reason for us not to systematically redescribe Aristotle's biological works using the term "matter" in places where he does not use this term. Aristotle almost always uses the term when it is relevant to the discussion that the thing is functioning *as* matter. When you describe something as matter out of context you can mistakenly ascribe features to it that do not hold insofar as it is matter.

There is a question remaining for this account: Does calling the (non-uniform) parts matter undermine the idea that matter is what something properly speaking comes to be from? After all, it is not clear that wholes come to be from their parts. This, in short, is Ackrill's classic puzzle about matter.[12]

Aristotle's view in the *Parts of Animals* is that things do, in fact, come to be from their parts. The generation of animals involves the successive generation of their parts until the whole organism is formed. Thus, here too matter is what a thing comes to be from. We can see that Aristotle is committed to such a view at the beginning of *PA* II.I. Let us examine this passage both (1) to finish our discussion of parts as matter, and (2) to turn

[11] For a similar account, see Code's chapter in this volume, section 2.

[12] Ackrill 1972/73.

to a new topic, the idea that matter is for the sake of what it becomes. Here is the passage:

> In generation things are opposed to the way they are in being (τῆς οὐσίας); for things posterior in generation are prior in nature, and the final stage in generation is primary in nature. For instance, a house is not for the sake of bricks and stones, but rather these are for the sake of the house – and so it is with other matter. Not only is it apparent from a consideration of cases that this is the way things are, but it also accords with our account; for every generated thing develops from something and into something, i.e., from an origin to an origin, from the primary mover which already has a certain nature to a certain shape or other such end. For a human being generates a human being, and a plant a plant, from the underlying matter of each. So the matter and the generation are necessarily prior in time, but in account the being and the shape of each thing. This would be clear if someone were to state the account of the generation of something; the account of housebuilding includes that of the house, while that of the house does not include that of housebuilding. And so it is in the other cases as well. Thus the matter of the elements is necessary for the sake of the uniform parts, since these are later in generation than the elements, and later than the uniform are the non-uniform; for these have already attained their end and limit, having achieved a constitution of the third sort, as often happens when generations are completed.
>
> Thus animals have been constituted from both of these parts, but the uniform parts are for the sake of the non-uniform . . . (646a24–b12)

As one would expect from the account of matter I provided above, Aristotle identifies matter as what things come to be from. He then infers from this that matter is prior in time to what it constitutes. One might wonder whether flesh really does develop prior in time to the parts constituted by flesh. But Aristotle's position in this passage is that it does. Given that he thinks of matter as what things come to be from, we can see why he would naturally think that it is prior in time to what it constitutes.

So far in this section we have seen no connection between matter and necessity in the *Parts of Animals*. The claim that matter is for the sake of something provides a further reason to be wary of reading a connection between matter and necessity. In some places, notably *GA* v, Aristotle claims that things can be necessary without being for the sake of anything. These are sometimes taken to be clear cases of material necessity.[13] But then, given what we have seen above, to precisely that extent they should not be due to matter, because matter *is* for the sake of something. One might want to say that iron is matter for the saw, but it is not for the sake of the

[13] For example, Leunissen 2010, p. 141.

saw. However, this is not Aristotle's view; in describing it as matter, we are describing it as for the sake of the saw. Aristotle says in the above passage that the bricks and stones are for the sake of the house, *and so it is with other matter*. Blood, in particular, is said to be for the sake of nourishment and growth. Thus, we should not think of any and all things explained non-teleologically as explained materially. On the contrary, Aristotle thinks that matter is teleological.

To summarize, on the few occasions when Aristotle mentions matter in providing his detailed accounts of animal parts, it is mostly identified as blood, because it is from this that the parts of animals nourish and grow. While other parts can be described as matter, this is not central to their function as a part. Matter is prior in time to what it becomes and for the sake of this thing.

Necessity

As mentioned earlier, in *PA* I.1 Aristotle says that we should look for two causes: the from-necessity and that for the sake of which (642a1–2). He goes on to do just this throughout the *Parts of Animals* and the *Generation of Animals*. The examples are ubiquitous.[14] In addition to identifying necessity as a cause, he uses it along with causal language, such as the use of *dia* with the accusative (because of). Cause (αἰτία) and related words show up 164 times in the *Parts of Animals*. Scholars often note that it is not clear how closely related *Parts of Animals* book I is to books II–IV.[15] This is one way in which the distinctive promises of book I are carried out in books II–IV. Someone who just read the *Physics*, say, or *Generation and Corruption* book I, would not be prepared for the focus in *PA* II–IV on "from necessity" and "that for the sake of which" as causes.[16]

This idea of "from necessity" as a cause seems rather strange. We know that Aristotle in many places, including the *Physics*, *Metaphysics*, and the beginning of *Generation of Animals*, says that there are four causes. And in a number of passages he suggests that that for the sake of which is the same as the form and the efficient cause. So this creates natural pressure toward thinking that "necessity" is a stand-in, of some sort, for matter. This seems

[14] The noun ἀνάγκη and the adjective ἀναγκαῖον show up 172 times in the *Parts of Animals*. Of course, not all of these uses are related to necessity as a cause.

[15] For example, Lennox 2001a. But Lennox 2010 argues for tighter connection.

[16] Contrast Leunissen 2010, section 4.3, where she treats Aristotle as giving formal and material causes throughout the *Parts of Animals* (on pp. 146–47 she identifies these as the primary causes). On my reading, Aristotle identifies the causes as from necessity and that for the sake of which; he occasionally mentions matter, but he is typically not giving matter as a cause of the feature to be explained.

to be one of the basic reasons that it is associated with matter.[17] There are other reasons as well, some of which will be touched on below.

My focus will be on providing a positive account of how Aristotle uses from-necessity as a cause. *Phys.* II.9 leads one to expect that the necessity in the *Parts of Animals* is a type of hypothetical necessity, where something is necessary given the hypothesis that some end is achieved. In fact, in a couple of places in *PA* I.I Aristotle suggests this same picture (639b21–640a9, 642a1–13). But, as is frequently pointed out, in most cases it is not this sort of hypothetical necessity that is identified as a cause in *Parts of Animals* II–IV. Instead, he treats certain features of animal parts as due to a more direct form of necessity. The last paragraph of *PA* I.I is the only place in the *Parts of Animals* where he seems to provide a general description of this other sort of necessity.[18]

> One should explain in the following way, e.g., breathing exists for the sake of *this*, while *that* comes to be from necessity because of *these*. But "necessity" sometimes signifies that if that – i.e., that for the sake of which – is to be, it is necessary for these things to obtain, while at other times it signifies that things are thus in respect of things possessed and naturally produced. For it is necessary for the hot to go out and enter again upon meeting resistance, and for the air to flow in. This is directly necessary; and it is as the internal heat retreats during the cooling of the external air that inhalation and exhalation occur. This then is the way of investigation, and it is in relation to these things and things such as these that one should grasp the causes. (642a31–b4)

Notice that we are told that this non-hypothetical sort of necessity comes from things "possessed and naturally produced." Assuming that this is what Aristotle pursues in the *Parts of Animals* books II–IV, we are left with the question of how and why "things possessed and naturally produced" (ἔχοντα καὶ πεφυκότα) necessitate. In particular, is this necessity directly connected to matter?

In *PA* I.I, Aristotle treats from-necessity explanations as something everyone agrees to, so he only seems to think that he needs to argue for the final cause. However, it is not clear that he accepts necessity as a cause for the same reason that Democritus or other Presocratics did. Thus, in order to

17 Monte Johnson in his discussion of *Po. An.* II.II (Johnson 2005, Chapter 2, section 3) and of Aristotle's biology (Johnson 2005, Chapter 6, sections 4 and 5) suggests that matter could be a specific case of the from-necessity cause (Johnson 2005). This is a complicated issue that I cannot adequately address here. My own view is that Aristotle may treat from-necessity and matter as different species of that-out-of-which, but matter is not a species of from-necessity, since it does not necessitate. This depends on some tricky issues in Aristotle's difficult *Po. An.* II.II.

18 Although, as Scharle points out in her contribution to this volume, at 640a23–b4 Aristotle suggests an important role for non-hypothetical, non-teleological necessity. Her account and discussion complement my own here.

understand how he thinks of necessity as a cause, it is useful to look at some of the actual explanations he gives in books II–IV. Necessity operates as a cause in a number of different ways in the *Parts of Animals*, so we will consider a number of passages to get an overview. Here is a simple case where both necessity and a final cause are cited and causal language is used, but in this case (unlike many others), the word "cause" is not used:

> With respect to the head, mankind is the most hairy of animals, from necessity, because of (διά) the moistness of the brain and because of (διά) the sutures (for where there is much moisture and heat there must be much growth), and for the sake of protection, so that it may provide covering, warding off the extremes of both cold and heat. And since the human brain is the most moist, it is also most in need of this protection; for what is moist boils and freezes most easily, while what is in the opposite state is less easily affected. (658b2–10)

The first question to ask about this sort of passage is why Aristotle thinks that it involves a type of necessity. The things described as necessary in the *Parts of Animals* are generally not strictly necessary, in the sense of impossible to be otherwise. To use the above example, it is not the case that human heads absolutely must be hairy. In Aristotle's language, this happens for the most part. There could be some sort of blockage in the sutures that does not allow the hair to grow. Moreover, as we know, Aristotle insists that there is no unqualified necessity in the sublunary realm.

These sorts of considerations lead Jacob Rosen to suggest that "necessity" in the *Parts of Animals* means something like compulsion or force – the sort of meaning it seems to have in Plato's *Timaeus*.[19] However, Lennox points out that this meaning of necessity as compulsion is treated as a special case in *Posterior Analytics* II.11, where Aristotle discusses necessity as a cause.[20] But Rosen is right to search for an explanation of why Aristotle thinks of these as cases of necessity. A different solution is suggested by *Generation of Animals*. There the idea is that something results of necessity when the agent and patient, identified as efficient cause and matter, come together. This is likely a shortened version of Aristotle's idea that when agent and patient come together, so long as nothing interferes, the result comes about of necessity. We find this idea, for example, in *Metaphysics* Theta 5 and *Phys.* VIII.4.[21] Perhaps when we discuss biology, we are discussing

[19] Rosen 2008, pp. 112–13. [20] Lennox 2001a, p. 149.

[21] This is not explicitly listed as a separate type of necessity in *Metaphysics* Δ 5, Aristotle's chapter on necessity. Perhaps it falls under the broader type of necessity that involves not being able to be otherwise. The focus in *GC* II.11 is on the guaranteed necessity of generation – a rather different topic.

how things normally are, in which case nothing is preventing. There are at least three passages in the *Generation of Animals* that suggest necessity resulting from this (731b18–23, 740b21–25, 778a34–b5). Although all three contribute something different, it is sufficient to consider one. One passage is in *GA* II.4; it is Aristotle's second explanation of why the uniform parts are formed:

> When the active thing and the passive thing, when they are in contact, are in the way that the one is acting and the other is being a patient (by "way" I mean the manner, the place, and the time), then immediately (εὐθὺς) the one acts and the other is passive. In this case, it is the female that provides the matter, and the male that provides the principle of movement. (740b21–25)

Aristotle's point is that in general when agent and patient come together in the right way, they straight away act; thus, this is how things will work in this particular case, where the mother provides the matter (and hence the patient) and the father the efficient cause (and hence the agent). In this passage Aristotle does not explicitly mention necessity, although he does in the other two passages (731b18–23 and 778a34–b5).

Aristotle's general principle, in *Metaphysics* Theta and *Physics* VIII, is put in terms of agent and patient. He often uses these terms interchangeably with moving cause (i.e., the efficient cause) and matter, although he is more likely to use the terms "moving cause" and "matter" when discussing generation and more likely to use the terms "agent" and "patient" when discussing alteration and locomotion.[22] Nonetheless, he does talk about the matter for alteration, locomotion, and growth and he does talk about the agent and patient in generation (as we saw in the previous passage).

There is a type of matter that is relevant to this sort of necessity. It is nonetheless misleading to call it material necessity because that suggests that matter plays a primary role in explaining these changes and results. Instead, the efficient cause is at least as responsible as the matter. Matter, taken on its own, is typically open to different possibilities. Blood can become different organs, or sustain them. It is only when blood is acted on in the right way that it becomes something specific. Moreover, it is worth understanding how Aristotle uses his own explanatory concepts. He almost never uses matter in his detailed explanations of animal parts. Rather, his goal – and what he does – is to trace things back to necessity as a cause. He is not trying to identify matter as a cause.

[22] See Code's contribution to this volume, section 2, for an argument that for Aristotle matter only strictly applies in cases of generation and corruption simpliciter.

The following is an example in the *Parts of Animals* of a from-necessity cause that fits this matter/efficient-cause model:

> The kidneys have the most fat of all the viscera. On the one hand, this is out of necessity, because the residue is filtered through the kidneys. For the remaining blood, being pure, is capable of good concoction; and soft and hard fat is an end of well-concocted blood. For just as in dry things that have been burnt, such as ash, some fire is left behind in them, so it is in moist things that have undergone concoction as well; that is, some portion of the heat which was operative is left behind in them. This is why what is oily is light and rises to the surface in liquids. Hence, on account of the visceral body being dense, the fat does not come to be in the kidneys themselves, but surrounds them on the outside, soft in the ones with soft fat, hard in the ones with hard fat. (The differences between these two sorts of fat have been stated previously elsewhere.) So on the one hand it is of necessity – this is the cause owing to which the kidneys come to be fatty, a consequence of what happens of necessity in animals with kidneys; on the other hand, they also come to be fatty for the sake of preservation of the kidneys and of their natural heat. (672a1–15)

Blood is matter for fat. Excess heat acts on the blood and turns it into fat; this process, of blood being heated, is how fat is normally formed (*PA* II.5). Although Aristotle does not do so, we could apply his concepts and say that the heat acts as an efficient cause on the blood, which is the matter. This happens of necessity, given the location of heat and blood. But it also is for the sake of preservation.

Not all cases of necessity fall under this matter/efficient-cause rubric. One example is important; it is the only case in the *Parts of Animals* where Aristotle explicitly mentions matter when giving an of-necessity explanation.

> The birds possess only auditory channels [rather than ears], on account of (διά) the hardness of their skin and because rather than having hair, they are feathered; accordingly they do not have the sort of matter from which ears may be formed. Likewise too with the four-footed animals that lay eggs and have hard scales – the same account also applies to these. (657a17–24)

While he does not explicitly say that birds lack ears from necessity, he gives the sort of explanation that elsewhere is identified as a from-necessity explanation: he explains why birds don't have something because it is impossible for them to have it, given some feature of them, in this case the hardness of their skin and their having feathers. But here, unlike elsewhere, Aristotle mentions matter. He says that accordingly (οὖν) they do not have the matter from which ears are formed. His point is that ears are formed from a certain sort of part, a certain type of skin, which birds lack. This skin is the matter for ears. When explaining why it is necessary that something *lacks* something else, you do not need both matter and an efficient cause. The absence of either matter or an efficient cause explains why the result cannot occur, since both are needed for the result to happen. This is the only way in which matter (on its own) can provide a necessity: a lack of matter can necessitate the lack of something.

This is related to a type of hypothetical necessity. Iron is hypothetically necessary for the saw and so, lacking iron, the saw cannot come to be. Similarly, soft, featherless skin is hypothetically necessary for ears and, lacking such skin, ears cannot come to be. In fact, such explanations are fairly common in the *Parts of Animals*. The cause of an animal lacking some feature is that it does not meet some necessary prerequisite for having the feature.

There are a number of cases of necessity in the *Parts of Animals* that cannot be explained in terms of matter and an efficient cause. We need to examine examples of these before we can pull together Aristotle's views about from-necessity as a cause. Consider this explanation of why birds are two-footed:

> They are two-footed of necessity; for the being (οὐσία) of the bird is that of the blooded animals, but at the same time that of the winged animals, and blooded animals do not move by more than four points. Accordingly, the attached parts are four – as in the other locomotive land-dwellers, so too in the birds. But four arms and legs are present in the one group, while in the birds, instead of forelimbs and arms, wings are a common feature; and in virtue of these they are able to stretch out, and the ability to fly is in the being (ἐν τῇ οὐσίᾳ) of the bird. So it remains for them to be, of necessity, two-footed; for in this way they will move, with their wings, by means of four points. (693b5–15)[23]

Birds have four limbs because they are blooded animals. But two of these need to be wings, because they are flyers. Thus, of necessity, they are two-footed. Notice that they are not two-footed for the sake of something. They have wings for the sake of something, but they are two-footed because of a combination of various features of their *ousia*. At the same time, these are direct features of their *ousia* that do not directly involve matter or an efficient cause.

Aristotle does not differentiate this sort of explanation from the earlier ones involving blood, heat, and skin: all are categorized as from necessity

[23] See also *PA* IV.9: a single row of suckers is of necessity given the *ousia* of a certain kind of octopus.

and are not further categorized or subdivided. Consider another case from *PA* II.1; this one comes shortly after the passage where he says that matter is temporally prior and for the sake of something. This example is unlike all of the others we have seen:

> As being for the sake of something, then – on account of *this* cause – these parts are related in the way stated [uniform parts constitute the non-uniform parts]; but when one also seeks how it is *necessary* that they be thus, it is apparent that they were antecedently so related to one another from necessity. For the non-uniform parts are capable of having been composed from the uniform parts, both from many of them and from one, as with some of the viscera; they are complex in configuration, though generally speaking they are composed of one uniform body. But it is impossible that the uniform bodies be composed from the non-uniform, for the uniform parts would consist of many non-uniform parts. These, then, are the causes owing to which some parts of animals are simple and uniform while others are composite and non-uniform. (646b27–647a2)

After arguing that the uniform parts are for the sake of the non-uniform, Aristotle provides this second argument, giving another cause of the one constituting the other. He offers what we might think of as a conceptual truth: non-uniform parts can consist of uniform parts but uniform parts cannot consist of non-uniform parts. You can compose a complex thing out of simple things, but you cannot compose a simple thing out of complex things. This explanation does not turn on matter, efficient cause, or the *ousia* –except, perhaps, *ousia* at a very high level of generality; it simply has to do with the very notion of a uniform and a non-uniform part.

From these cases, we can conclude that Aristotle does not think there is a single sort of necessity involved in all of these different types of "from-necessity" explanations: it is any necessity that is the result of things "possessed and produced naturally," as he says in the *PA* I.1 passage. He understands "possessed and naturally produced" quite broadly. Aristotle is content to bring things to necessity and then consider his job done. Why does he not think that to provide a full explanation, he needs to categorize the type of necessity at issue, or to bring things back to the four causes?

Here is one explanation we might be tempted to give: perhaps teleological explanations are important because they are not necessary. Thus, the epiglottis exists *for the sake of* not choking, but it is not necessary in order to not choke (664b20–665a8). There is another way to avoid choking: some animals' windpipes collapse. If teleological explanation were, in general, not necessary, this would be a good reason for Aristotle to treat these as the two types of explanations. The main problem with this idea is that some cases of for-the-sake-of causes – the ones that Mariska Leunissen calls primary teleology – are cases where something is both necessary and for the sake of something: it is necessary because it is part of the essence or vitality of the organism.[24] Birds have wings for the sake of flying and it is necessary for birds to have wings, given that they are essentially flyers.[25] Moreover, as we have seen, in some cases – such as the fat around the kidney – something is both for the sake of something and, for a different reason, necessary.

Here is another explanation that might be tempting: perhaps Aristotle is primarily interested in what each of the parts is and to discover that we need to know what each is for. Our central task is to determine what it is, and everything else is secondary and so is shelved under the category "necessary." The problem with this explanation is that some necessary features result from what a thing is. If we want to know what birds are, we need to understand that "of necessity" they are two-footed, even though this is not for anything.

Given the problems with those explanations, the most likely account is the following: Aristotle's goal is to bring out the importance of teleological explanations. He wants to show that, in addition to those other explanations people give, in terms of necessity, you must also give teleological explanations (this is a theme throughout *PA* I.1). Aristotle accepts his predecessors' reliance on necessity as a cause (as he understands his predecessors) and sees himself as adding to their type of account.

The *Parts of Animals* might also be, in one respect, closer to a more Platonic way of thinking about the natural world. It is striking that Aristotle's two causes seem closely related to the explanatory structure of Plato's *Timaeus*, where the Demiurge does everything for the sake of what is best (29e–30a) and necessity is introduced as an additional part of the explanatory framework and called the wandering cause (47e–48b). There are a number of very important differences between Aristotle's explanatory project and Plato's. But perhaps Aristotle's two causes are more close to the *Timaeus* than Aristotle's account in other works. I have argued elsewhere that Aristotle sees his four causes as anti-Platonic.[26] Perhaps *Parts of Animals* II–IV was composed before he took on this more anti-Platonic account of causation. It is easy to imagine that it was composed while in Assos or Lesbos – although that is pure speculation. The other possibility is that he

[24] Leunissen 2010 *passim*, first explained at p. 4 and pp. 18–19.
[25] To be clear, in such cases necessity is not listed as a cause; the point is just that being for the sake of something is not an alternative to being necessary.
[26] Ebrey 2014.

thinks, for some reason, that in the *Parts of Animals* we only need these two causes, despite the fact that in the *Phys.* II.7 he insists that the natural scientist identify all four causes for each thing investigated (198a21–24).

Conclusion

I have argued that the role of matter in the *Parts of Animals* is quite different from that of necessity. When Aristotle mentions matter in *Parts of Animals* II–IV, he reserves the term for what things properly come to be from. This is generally identified as blood, the thing that the body concocts as the refined nutriment, able to become any of the parts or to fuel their growth and nourishment. By contrast, Aristotle frequently identifies "of necessity" as a cause. In many cases this involves what we might identify as the matter and efficient cause working together, although Aristotle never identifies them this way in the *Parts of Animals*. Other examples of "of-necessity" causes involve facts about the essence of the creature or some sort of conceptual necessity. Aristotle is not interested in separating out the different types of necessity at work in these examples. Instead, he draws everything back to the necessary and that for the sake of which. Matter and necessity have very different roles in the *Parts of Animals*.

PART II

Teleology

CHAPTER 5

"And these things follow"
Teleology, necessity, and explanation in Aristotle's Meteorologica

Margaret Scharle

Aristotle's *Meteorologica* seems to be completely free of teleology, offering solely mechanistic explanations for meteorological phenomena.[1] In fact, Aristotle uses no teleological language in the entire treatise. In this chapter, I argue that there is nonetheless an important explanatory role for teleological processes to play in the treatise, and I detail this role by considering analogous processes in the biological works.

Section I offers prima facie evidence that the explanations found in the *Meteorologica* should fit into *de Partibus Animalium*'s model of explanation in natural science. Section II begins to fill out the model of explanation by showing that the ultimate *explanantia* in meteorology – the elemental cycles of generation – are the inorganic analogues of the ultimate *explanantia* in biology – the organic generative processes including "man generates man." Not only do the inorganic and organic generative processes play analogous roles as ultimate *explanantia* within their subfields of natural science, but furthermore when they are viewed as *explananda* from the cosmic perspective taken up in the *Metaphysics*, we can observe their common causal etiology grounded in their relationship to the prime mover. Section III, nonetheless, reveals an important disanalogy between the inorganic and organic generative processes that plays out in the way the inorganic and organic subfields of natural science fill out the model of explanation in natural science. This difference explains why the biological works such as *de Partibus Animalium* use teleological language, while the *Meteorologica* does not. Finally, Section IV shows the explanations found in the *Meteorologica* to fit the inorganic schema of natural scientific explanation suggested in Section III, thereby explaining why teleological processes play an

I thank the participants in the Northwestern Theory and Practice workshop, and especially David Ebrey for his written comments on the penultimate draft. I also thank Kellyn Bardeen for her excellent editorial assistance and philosophical insight.

1 For a recent example of this interpretation, see Wilson (2014, Ch. 5).

important role in meteorological explanation despite the fact that no teleological language is used in the treatise.

Section I Three kinds of explanation in natural science

This section offers prima facie evidence that the explanations found in the *Meteorologica* fit *de Partibus Animalium*'s model of explanation in natural science. Not only does the *Meteorologica* locate itself within natural science and claim to share a method (τὸν τρόπον; 339a7) with the biological works, but *de Partibus Animalium* claims its method (τὸν τρόπον; 640b3–4) of natural scientific explanation should apply to all natural things (πάντων τῶν φύσει συνισταμένων; 640b4). This would include the natural elements (*Phys.* II.1, 192b9–11) whose generations and movements are the subject of the *Meteorologica.* With this prima facie evidence in hand, the rest of my chapter argues that the model of natural scientific explanation set out in *de Partibus Animalium* is indeed the model followed in the *Meteorologica.*

Let us begin with the *Meteorologica.* Situated squarely within the science of nature – commenced in *Physics*, followed by *de Caelo*, *de Generatione et Corruptione*, and carried through the biological works – the *Meteorologica* concerns *ordered*, *natural* phenomena, albeit ones that fall short of the order of the heavenly bodies. Aristotle highlights the unity among all the natural scientific investigations, and asserts, in particular, that the method (τὸν τρόπον; 339a7) followed in the *Meteorologica* will be used in the biological works:

> We have already discussed [in *Phys.*] the first causes of nature, and all natural motion, also [in *DC* I and II] the stars ordered in the motion of the heavens, and [in *DC* III and IV and *GC*] the corporeal elements – enumerating and specifying them and showing how they change into one another – and becoming and perishing in general. There remains for consideration a part of this inquiry which all our predecessors called meteorology. It is concerned with events that are natural (κατὰ φύσιν), though their order is less perfect than (ἀτακτοτέραν) that of the first of the elements of bodies . . . When the inquiry into these matters is concluded let us consider what account we can give, in accordance with the method (τρόπον) we have followed, of animals and plants, both generally and in detail. When that has been done we may say that the whole of our original undertaking will have been carried out (τῆς ἐξ ἀρχῆς ἡμῖν προαιρέσεως πάσης) (I.1, 338a20–b4, 339a6–9).[2]

[2] All translations are from Barnes (1984), unless otherwise noted.

Here, Aristotle clearly indicates his "method" (τρόπον) is applied in consistent manner to all the subjects of natural science. Since *de Partibus Animalium* illuminates Aristotle's approach to natural scientific subjects in greater detail than the *Meteorologica*, we will examine this text to gain a greater understanding of the particularities of his method. *de Partibus Animalium* explicitly states that its explanatory method (τὸν τρόπον; 640b3–4) applies to *all natural things*, reassuring us that this method applies to the investigation of meteorological phenomena as well. Aristotle's general approach to natural science is detailed in the following passage from *de Partibus Animalium* I.I. Beginning with the claim that in nature, as in housebuilding, the coming to be of parts should be explained by reference to the form that is being realized, Aristotle continues:

> [P]ropagation implies a creative seed endowed with certain powers . . . For (γὰρ) man is generated from man; and thus (ὥστε) it is because the parent is such and such that the generation of the child is thus and so . . . Hence [*a*] it would be best to say that, since (ἐπειδὴ) this is what it is to be a human being, on account of (διὰ) this it has these things; for it cannot be without these parts. [*b*] If one cannot say this, one should say the next best thing, i.e. either that in general it cannot be otherwise, or that at least it is good thus. [*c*] And these things follow (ταῦτα δ᾽ ἕπεται). And since it is such, its generation necessarily happens in this way and is such as it is. (This is why this part comes to be first, then that one.) And in like manner one should speak in precisely this way about all of the things constituted by nature (καὶ τοῦτον δὴ τὸν τρόπον ὁμοίως ἐπὶ πάντων τῶν φύσει συνισταμένων). (*PA* I.I, 640a23–b4)[3]

The approach is teleological: the empirical fact that "man generates man" *implies* (signaled by his use of "for" [γὰρ] and "thus" [ὥστε] [640a25–26]) that the (male) parent animal has a creative seed with the power to produce an offspring of the same form. The parts of humans are explained teleologically by reference to this form (signaled by the use of "since" [ἐπειδὴ] and "on account of" [διὰ] [640a34–35]); thus, Aristotle's teleological explanation of the parts of animals takes the fact that "man is generated from man" as the ultimate *explanans*: given that man generates man, we can understand every step of that generative process by reference to the fact that another man is being generated.

Aristotle's approach identifies three different forms of explanation, marked in the passage as [*a*], [*b*], and [*c*]. Drawing on examples throughout *de Partibus Animalium*, Mariska Leunissen illuminates the nature of the

[3] Trans. Lennox (2001a), modified.

explanations distinguished here in I.1.[4] "Primary teleological explanation" (labeled as [*a*] in the passage) explains the presence of a part as being necessary for there being something of that form ("it cannot be without these parts"). Paradigmatic examples include the heart and liver, which are vital for blooded animals (*PA* II.1, 647a35–b8; III.4, 665b10–15; IV.3, 677a36–b5), and fins, which are essential to fish, given that it is "in accordance with their essential substance they are swimming animals . . . and since they are made for swimming they have fins" (*PA* IV.13, 695b17–26).[5] What she calls "secondary teleological explanation" ([*b*] in the passage) is that which explains the presence of "subsidiary" parts, which are necessary ("cannot be otherwise") and support the functioning of the primary parts (e.g., the kidneys support the bladder [*PA* III.7, 670b23–27]), and of "luxury" parts such as nails, hair, hoofs, horns, beaks, and spurs of cocks (*GA* II.6, 745a1–2), which are neither primary parts nor support the functioning of the primary parts, but are "at least . . . good thus" insofar as they contribute to the well-being of the animal.

Drawing on *de Generatione Animalium* II.6, Leunissen argues that the primary parts are made first out of the best nutriment, the secondary subsidiary parts are made second out of the inferior nutriment (745a3), and the secondary luxury parts are made third out of the residues left over from the process of making these other parts.[6] The formal nature of the animal, like a good housekeeper, "is not in the habit of throwing away anything from which it is possible to make anything useful" (744b16–17), and makes the best use of the nutriment as well as the leftovers from the part-making processes. Nonetheless, the formal nature cannot make use of every bit of material, and dregs remain after the formal nature has picked through the leftovers. As necessary by-products of the formation of the primary and secondary parts, these dregs have no teleological explanation. All Aristotle can offer by way of explanation is that "these things follow (ταῦτα δ᾽ ἕπεται)" ([*c*] in the passage), which Leunissen thinks "could refer to a third form of explanation in which references to teleology are completely absent."[7]

I think we should be confident that this is indeed a third form of explanation, since it alone fits Aristotle's explanation of biological by-products, such as the spleen, bile, and residues of the gut and bladder (III.7, 670a30 ff), but this is where I will part from Leunissen's theory. Aristotle's explanations concerning these by-products support the claim that "and

[4] Leunissen (2010, 81–99).
[5] These and other examples are cited by Leunissen (2010, 85 ff).
[6] Leunissen (2010, 81–84).
[7] Leunissen (2010, 98).

these things follow" explanations are not *themselves* teleological, and yet, the text casts doubt on Leunissen's suggestion that such explanations do not even *reference* teleology. On my interpretation, "and these things follow" explanations *must* refer to teleology: These things follow . . . what precisely? They follow from teleological processes. Bile is a case in point:

> When [bile] is present in the region of the liver it is a residue and not for the sake of anything; just as is the case with the excretions of the stomach and intestines. For though even the residua are occasionally used by nature for some useful purpose . . . we must not in all cases expect to find such a final cause; for granted the existence of this or that constituent, with such and such properties, many results must ensue as necessary consequences of these properties. (*PA* IV.2, 677a14–19)

After explicitly denying that bile has a teleological explanation, because the formal nature can make no "useful purpose" of such dregs, Aristotle goes on to *reference* the teleological functioning of the liver in order to explain the location and production of bile. Given that blood is sometimes "less pure in composition," there must be an organ in blooded animals to purify the blood by excreting bile (677a26). There are only two vital organs that all blooded animals share – the liver and heart – and the heart is "unable to stand any violent affection [and] would be entirely intolerant of the proximity of such a fluid," so Aristotle concludes that the liver is the organ to serve this purification function ("none other should have an excretion as the bile") (*PA* IV.3, 677a19–b10). Thus bile is where it is (next to the liver, instead of next to the heart) because the liver serves the function of purifying the blood. And the liver, as a primary part, serves this function so that the animal can live. Indeed the animal's length of life depends on the well-functioning of the liver (677b1–2). Thus, Aristotle's explanation of bile seems to fit the "and these things follow" form of explanation: although bile does not itself have a function, its production and location are explained by an indispensable reference to the liver's teleological operations.

As indicated above, the significance of this explanatory approach extends beyond *de Partibus Animalium.* Although the discussion above takes as its immediate concern *biological* explanations, Aristotle concludes the passage by insisting that "in like manner one should speak in precisely this way about all of the things constituted by nature (καὶ τοῦτον δὴ τὸν τρόπον ὁμοίως ἐπὶ πάντων τῶν φύσει συνισταμένων)" (*PA* 640b3–4).[8] Given that these explanations apply to *all natural things,* I therefore propose naming these "three forms of natural scientific explanation" to signal their intended

[8] Trans. Lennox (2001a).

broad scope. Prima facie, then, they provide the explanatory schema for the *explananda* of all the natural scientific treatises. In what follows I show the *Meteorologica* to be such a treatise.

Section II "Man generates man" and its meteorological analogues

The previous section provides at least prima facie evidence that the explanations found in the *Meteorologica* should fit into *de Partibus Animalium*'s model of explanation. This section will begin to fill out the explanatory schema for the *Meteorologica* by arguing that the elemental cycles of generation are the analogues of "man generates man." The analogy can be drawn from two perspectives. First, within the natural scientific subfield of meteorology, the elemental cycles are introduced as ultimate *explanantia*, just as within the natural scientific subfield of biology organic generative processes (such as "man generates man") serve as the ultimate *explanantia*. Second, from the cosmic perspective adopted by texts such as the *Metaphysics*, the inorganic (the elemental cycles) and organic generative processes (such as "man generates man") come into view as *explananda* that take the prime mover as their efficient and final cause. I thereby suggest that the ultimate *explanantia* in the subfields of natural science are the teleological, generative processes that are themselves explained by their relation to the prime mover, a relationship we comprehend only from the cosmic perspective.

Let us begin with the first perspective from within the subfield of the natural science of meteorology. The cycles of elemental transformation (or in my terminology, "originative cycles") are introduced when Aristotle takes up his first subjects of meteorological explanation in I.4:

> The origin (ἀρχή) of these and many other phenomena is this. When the sun warms the earth the exhalation (ἀναθυμίασιν) which takes place is necessarily of two kinds, not of one only as some think... That which rises from the moisture contained in the earth and on its surface is vapour (ἀτμίδα), while that rising from the earth itself, which is dry, is like smoke (καπνώδη)... [T]he windy exhalation, being warm, rises above the moister vapour, which is heavy and sinks below the other.
>
> Hence the world surrounding the earth is ordered as follows. First below the circular motion comes the warm and dry element, which we call fire, for there is no word fully adequate to every state of the smoky evaporation; but we must use this terminology since this element is the most inflammable of all bodies. Below this comes air. We must think of what we just called fire as being spread round the terrestrial sphere on the outside like a kind of fuel (ὑπέκκαυμα), so that a little motion often makes it burst into flame

> just as smoke does; for flame (φλόξ) is the ebullition of a dry exhalation. So whenever the circular motion stirs this stuff (σύστασις) up in any way, it catches fire at the point at which it is most inflammable. (I.4 341b6–24)

In this passage, Aristotle seeks to identify the origin (ἀρχή) of meteorological phenomena in two "originative" cycles of elemental transformation. The sun warms the earth (cold and dry) and transforms it into a dry exhalation (hot and dry), which rises to the top and is ignited by the movement of the celestial bodies. As I will argue, earth's transformation into fire can be seen to compose one leg of an earth–fire cycle. But, more clearly, Aristotle understands air and water to form an originative cycle: the summer sun warms the water (cold and wet) and transforms it into a wet exhalation (hot and wet). When the sun recedes in the winter and the exhalation is thereby cooled, it transforms back into water (cold and wet) and falls back to the ground, thereby forming a water–air cycle:

> For the sun as it approaches or recedes, obviously causes dissipation and condensation and so gives rise to generation and destruction . . . So we get a circular process that follows the course of the sun. For according as the sun moves to this side or that, the moisture in this process rises or falls. We must think of it as a river flowing up and down in a circle made up partly of air, partly of water. When the sun is near, the stream of vapour flows upwards; when it recedes, the stream of water flows down; the order of sequence, at all events, in this process always remains the same. (I.9, 346b21–347a6)

The other originative cycle of earth–dry exhalation–fire (hereafter called simply the "earth–fire cycle") is less obviously cyclical. Aristotle only details one leg (the upward direction) of this originative cycle: the sun transforms earth into dry exhalation that rises to the top and bursts into flames by the friction caused by the superlunary motion. Given that the sun and other heavenly bodies regulate these transformations, the process is regular: the dry exhalation is "continually and ceaselessly (συνεχῶς καὶ ἀεὶ)" collecting in the upper region (I.8, 346a23) and "at each revolution of the heavens (καθ' ἑκάστην περίοδον)," it is ignited (I.8, 346b8–10). Although Aristotle does not focus on the downward progression of this cycle (fire to earth), there is at least some evidence that he thinks this downward direction exists: in II.3 he includes fire, along with air and water, as that which is "involved in a process of perishing and becoming which yet affects them all in a regular course" (II.3, 357b27–358a7). In the case of water and air, Aristotle is certain that "all that is carried up is returned" (355a23–24) since "we always plainly see the water that has been carried up coming down again" (355a25–26). So perhaps Aristotle does not detail the downward leg of the earth–fire cycle since, unlike in the case of the water–air cycle, we do not have observational data to rely on. Perhaps the downward leg is not a meteorological phenomenon.

But neither is a defense of the downward leg of the cycle dialectically required. Aristotle details one cycle of water–air in response to two groups of predecessors who shared a non-cyclical, linear conception of elemental change: the first group thought that rivers and winds issued forth from a reservoir of ready-made material, and the second group, the entropists, thought that the earth was undergoing an irreversible process of desiccation. By contrast, when Aristotle describes the process of earth's transformation into fire, instead of arguing against predecessors who insist the process is linear, here, as Malcolm Wilson points out,[9] his target is Anaxagoras. Anaxagoras' theory of meteors contradicts the paradigm example in *Phys.* II.1 of natural elemental movement – that of fire moving upward (193a1). To preserve the natural movement of fire upward in the face of fire's apparent horizontal movement in the form of meteors and comets, Aristotle argues that the apparent horizontal movement of a comet is actually like fire sweeping across a field of stubble, the sequential production and ignition of dense heaps of exhalation by the moving planets and stars. Hence, the details of the downward arm of the earth–fire cycle are not dialectically required, because the importance of the question at hand is not the cyclical nature of the process. Although more could be said, I will regard it as established that there are at least two originative cycles within Aristotle's account of the origins of meteorological phenomena, and will refer to the plural, "originative cycles."

The movement of the sun is said to regulate both of the cycles I have described. The claim in *Meteorologica* I.2 that "this world" derives "all its power" from the superlunary motions (339a20–33) applies this point more broadly, and refers to the claim in *Physics* VIII that the prime mover is the ultimate efficient cause of all things. Identifying the sun as the immediate efficient cause of elemental transformation, Aristotle shows that the orderliness of the superlunary processes is translated down via the circle of the sun into the regularity of the sublunary originative cycles: "The efficient and chief and first of the principles is the circle in which the sun moves" (I.9, 346b20–21).

The *Meteorologica* affords us only a glimpse of the cosmic perspective that brings into view the causal relationship the originative cycles bear to the prime mover, for the task of the *Meteorologica* is to explain phenomena

[9] Wilson (2014).

by reference to the originative cycles, not to explain the originative cycles' causal relationship to the prime mover. That is, the text introduces the originative cycles as ultimate *explanantia*, not as *explananda*. Likewise, as we have seen, "man generates man" is introduced in *de Partibus Animalium* as the ultimate *explanans*, not as an *explanandum*. The cosmic perspective adopted in texts outside the *Meteorologica* and *de Partibus Animalium* affords us a view of the generative cycles as *explananda* themselves: from the cosmic perspective we see that the non-biological originative cycles share an efficient and final causal etiology with biological generative processes such as "man generates man."

Just as *Meteorologica* I highlights the fact that the originative processes are efficiently caused by regular movement of the sun along the ecliptic, *de Generatione et Corruptione* identifies the sun as the efficient cause of "man generates man," a point Aristotle highlights several times throughout his corpus: "since the upper movement is cyclical, the sun moves in this determinate manner; and since the sun moves thus, the seasons (ὧραι) in consequence come-to-be in a cycle, i.e. return upon themselves; and since they come-to-be cyclically, so in their turn do the things whose coming to be the seasons initiate [e.g., plants and animals]" (*GC* II.11, 338b3–5; see also *DC* II.3, 286a13–b2). This is the sense in which "man is begotten by man and by the sun as well" (*Phys.* II 2, 194b13): the sun is the efficient cause of the seasons, and the seasons provide living things with nutrients required for generation and growth.

It is clear that the originative cycles and the generation of man share the sun as their efficient cause. But while the argument in *Meteor.* I.4 is concerned with this efficient causal dependence on the movement of the sun, and ultimately on the activity of the prime mover, we should not mistake this focus to exclude the possibility that the originative cycles are also teleological. The cosmic perspective shows that just as "man generates man" takes the sun, and ultimately the prime mover, as its (remote) efficient causes, it also takes the prime mover as its *final* cause. These same texts suggest that the originative cycles share this final causal structure with their biological analogues: the prime mover is the final cause of *both* the biological and the non-biological generative processes. In what follows, I review select passages that illuminate this point.

According to the conception of natural teleology in *Physics* II, nature acts for something and because it is better (II.8, 198b17–18). Given that plants, animals, and the elements have a nature (*Phys.* II.1, 192b9–11), they have teleological direction. Although artifacts such as beds and houses enjoy a teleology afforded to them by their creators (e.g., beds are for supporting sleepers), all natural things exhibit a distinctive teleology at the level of generated wholes, a distinction Aristotle marks in his argument in *Phys.* II.1 against materialist predecessors such as Empedocles: "man is born from man but not bed from bed" (193b9–10). Thus, the teleology distinctive of natural things is that of continuous generation, for nature is distinguished from art by an *internal* principle of "production (ποιήσεως)" (*Phys.* II.1, 192b30).

The teleology of continuous generation is focused on imitating the highest good in the cosmos, the prime mover, as the "one thing" to which everything in the cosmos bears a relation, as the cosmic perspective of *Met.* XII.10 reveals (1075a17–18). Individuals strive to *be* the prime mover, which is eternal, purely noetic activity. But since they can never successfully achieve this aim, the most they can do is *approximate* it through imitation. The heavenly spheres directly imitate the prime mover's activity by eternally moving in perfect circles; since each point on a circle is as much an end as any other point, circular motion imitates the self-ended activity of the prime mover (*Phys.* VIII.9, 265a28–b9, referenced in *Meteor.* I.2, 339a25). Plants and animals imitate the eternal actuality of the prime mover by reproducing another of their kind (*DA* II.4, 415a25–b7), while the sublunary elements indirectly imitate the prime mover by being part of a cycle that imitates the superlunary circular movements: "we say coming to be has completed the circle, because it reverts again to the beginning. Hence (ὥστε) it is by imitating circular motion that rectilinear motion too is continuous" (*GC* II.10, 337a5–7).[10] Thus, rectilinear movement is imitative only when and *because* (expressed by ὥστε) it occurs on the heels of another rectilinear movement with which it composes a cycle.

Although the relationship that the sublunary generative processes bear to the prime mover is comprehended from the cosmic perspective of the *Metaphysics*, the *Meteorologica* itself emphasizes the naturalness and regularity that the imitative teleology of *Met.* XII.10 requires. For example, the diurnal and annual cycle of evaporation and condensation are described as the regular course of nature – φύσις (II.2, 354b34), κατά ἱ τὴν τάξιν (I.9, 347a6), κατά τινα τάξιν, ὡς ἐνδέχεται μετέχειν τὰ ἐνταῦθα τάξεως (II.3, 358a26–27), περιόδου (I.14 352a31), and τεταγμένως (II.3 358a3, quoted above). In addition to the diurnal and annual cycle of evaporation and condensation, Aristotle recognizes a third form of the water–air cycle: "in determined intervals in some great period of time there comes a great winter (μέγας χειμὼν) and with it great excess of rain" (352a29–33;

[10] See also *GC* II.10, 336b25 and 336b32–337a1.

see also II.3 357a2), which is counter-balanced by periods of drought. Once again, the cycle is regular – περιόδου (I.14 352a31), τεταγμένοις (II.2, 355a28; II.3, 358a3, quoted above), κατὰ κύκλον (II.3, 357a2), κατὰ ἱ τινὰ τάξιν ἱ καὶ περίοδον (I.14, 351a26 and 352b16) – and natural (φυσικὴν I.14, 351b8–14; see also II.3, 359b23–26). This emphasis on naturalness and regularity suggests that the imitative teleology of *Met.* XII.10 is at play in the background.

Thus, I have argued that the cosmic perspective shows us that the biological and non-biological generative processes are *explananda* with the same final and efficient explanations, an etiology reflected in the *Meteorologica*'s emphasis on the originative cycles' naturalness and regularity. And just as *de Partibus Animalium* shows "man generates man" as the ultimate *explanans* within the natural scientific subfield of biology, we have seen the originative cycles introduced as the ultimate *explanantia* in the natural scientific subfield of meteorology. The suggestion, then, is that the processes that serve as ultimate *explanantia* within a subfield of natural science are the generative processes explained by their teleological relationship to the prime mover from the cosmic perspective.

Section III The disanalogy between inorganic and organic generative processes

Section II argued that the biological and non-biological processes of generation are the same sort of *explananda* from the cosmic perspective and, in turn, play an analogous role as ultimate *explanantia* within their respective subfields of natural science. This section points out an important disanalogy between organic and inorganic generative processes. I first illuminate the disanalogy from the cosmic perspective and then return to the subfields of biology and meteorology to show how the disanalogy plays out in the particular ways each subfield fills out the model of natural scientific explanation examined in Section I.

Let us begin, then, with the cosmic perspective. As we have seen, all sublunary generation aims to imitate the activity of the prime mover; however, Aristotle insists that the greater and lesser extents to which individuals partake in the divine creates a hierarchy of beings stretching from the heavenly bodies – which are "more divine" (I.2, 269a32–33) since they have "a higher nature" (I.2, 269b16–17) – all the way down to the sublunary elements, which *de Caelo* II.12 confirms, "share in the divine source (τυχεῖν τῆς θειοτάτης ἀρχῆς)" (292b17–25). *De Generatione Animalium* II.1 corroborates this ranking: beginning at the top of the *scala naturae*, Aristotle places the heavenly bodies (both living and eternal), then living things (living but not eternal), and finally the sublunary elements (neither living nor eternal) (731b24–732a1). The household analogy of *Met.* XII.10 alternatively calibrates the hierarchy according to regularity: the heavenly bodies are to the sublunary elements as the freemen are to the slaves and beasts insofar as the heavenly bodies "have least license to act as they chance to, but all or most of what they do is arranged (τέτακται)," while the sublunary elements "can do a little towards what is communal, but act mostly as they chance to" (1075a19–22; see also *DA* II.4, 415b4–6).[11] The introduction to the *Meteorologica* confirms this contrast in regularity: the sublunary elements exhibit a "regularity less (ἀτακτοτέραν) than" the heavenly bodies moving in a circle (*Meteor.* I.1, 338a20–b4). Calibrating the *scala naturae* according to regularity comes to the same thing as calibrating it according to divinity or "goodness": since the prime mover is the best thing in the cosmos, the more closely something approximates its activity, the better it is; and the more closely it approximates the prime mover, the more regular and uninterrupted its activity.

This difference in degree of regularity exhibited by the generation of elements and living things can be explained by the difference in formal natures involved in each generative process. *Phys.* II.1 defines nature as a "source of movement" (ἀρχὴ κινήσεως), and maintains that both living things and sublunary elements have a nature (II.1, 192b9–11). However, *Phys.* VIII.4 goes on to suggest that "movement" (κινήσεως) is ambiguous between the active (κινεῖν) and passive (κινεῖσθαι) senses. While self-moving living things have a source of moving (an ἀρχὴ τοῦ κινεῖν), sublunary elements' nature is a source of suffering (an ἀρχὴ τοῦ πάσχειν 255b31), which can be understood as a source of being moved (ἀρχὴ τοῦ κινεῖσθαι), in keeping with the II.1 definition.

Phys. VIII.4 further maintains that the elements are passive not only with respect to their transformation into one another, but also with respect to their movement into natural place. Their generator is responsible for both their generation and their movement. Aristotle arrives at this claim in the course of arguing for the conclusion that everything that moves is moved by something (255a1–6). Initially puzzled by what moves the elements since they have no parts and thus cannot move themselves, Aristotle traces the confusion to the failure to recognize that "potentiality is said in various ways" (255a30). Once we distinguish the two sorts of potentiality – the second potential to move into natural place and the first potential to come

[11] Trans. Sedley (2000, 328).

to be – we see that movement is the *second* phase of a single two-phase process (of which coming to be is the first stage). Once we realize that the two phases compose a single process, it is clear that the generator is the efficient cause of elemental movement into natural place: the chapter concludes, "[t]he light and the heavy . . . [are moved] by the thing that has generated and has made them light or heavy" (256a1–3). While we had trouble identifying the efficient cause of elemental movement into natural place (the second phase), it is clear that the efficient cause of the first phase is the generator of the elements. Although Aristotle does not mention what this generator is (since all that matters for his purposes here is that there is one), the *Meteorologica* is clear that at least *a* generator is the sun moving along the ecliptic (I.9, 346b22 and *GC* II.10, 336b6–7).

Although our analysis at the beginning of this section suggested that living things and elements share an efficient causal etiology traced back to the sun as their generator, this *Phys.* VIII.4 distinction between active and passive natures explains the difference registered by their respective places on the *scala naturae*, where the living ranks above the not living (*GA* II.1, 731b24–732a1). As we saw in Section 1, *de Generatione Animalium* II.6 personifies the formal nature of a living thing as a good housekeeper that does its best to make use of the materials at hand. Although the offspring takes the father, and (remotely) the sun, as its generator, the offspring's active formal nature continues to make use of nutriment and leftovers to form and grow parts even after birth and into old age (*GA* II.6, 744b31–745a18). Making use of such materials, an animal can not only live, but live well and ever better, thereby actualizing its nature to higher degrees, the apex of which is generating another of its own kind in imitation of the prime mover. Living things' active natures are, moreover, able to refashion their initial by-products into teleologically directed subsidiary and luxury parts. For example, hair comes to be as a materially necessitated by-product of the development of vital organs. Instead of letting that hair go to waste, the organism actively co-opts the hair for the sake of protection against heat and cold.[12] This refashioning requires the kind of active nature unique to self-moving living things.

By contrast, the elements are utterly passive with regard to the timing and location of their generation and are unable even to stop themselves from moving (*Phys.* VIII.4, 255a8–11). The elements cannot have the active nature of self-movers precisely because they have no parts: "[N]one of these things move themselves – for they are naturally unified (συμφυῆ) – nor does

[12] See Leunissen's description of secondary teleology (2010, 81–99).

anything else that is continuous" (255a15–16). Insofar as the elements enjoy a more immediate relation to the superlunary order of the sun than living things do, the elements' passivity is an asset: the sun's orderly approach and retreat ensures that air's upward rectilinear movement will be preceded and followed by water's downward rectilinear movement to form an imitative cycle (*GC* II.10, 337a5–7; *Meteor.* I.9, 346b16–347a6), while the sun's orderly movement has a more indirect influence on living things insofar as it must be mediated through organisms' use of the food that the seasons provide (*GC* II.11, 338b3–5). However, much of the meteorological phenomena that occur simply follow from the originative cycles as by-products, which the passive natures of the elements are unable to refashion into anything useful for the purpose of imitating of the prime mover.

Now we are in a position to see how this difference in active and passive natures, rooted in the difference between being an entity composed of parts and an element that is a partless whole, impacts the kinds of explanations that can be given in the biological and non-biological subfields of natural science. Recall the three forms of natural scientific explanation as they apply to organic things: man generates man (its teleology in view only when seen as itself an *explanandum* from the cosmic perspective), therefore he has (i) these vital and necessary parts (primary teleological explanation), and (ii) these subsidiary and luxury parts (secondary teleological explanation), and (iii) "and these things follow" (the third, non-teleological explanation). But as utterly passive and without parts, the elements have no parts to be explained by primary and secondary teleology, and thus the explanatory schema can have no analogue of the first and second sort of explanation, but could only have an analogue of the third, filling out the explanatory schema as follows: originative cycles (its teleology in view only when seen as *explananda* from the cosmic perspective) "and these things follow" (the third, non-teleological explanation). Given that the originative cycles are themselves teleological, we can make sense of Section I's suggestion that although "and these things follow" explanations are not *themselves* teleological, these explanations nonetheless *must* refer to teleology: These things follow . . . what precisely? They follow from teleological processes. As teleological processes, the originative cycles can explain the phenomena that result from them, even though those phenomena are not themselves teleologically directed. At the same time, filling out the schema in this way would make sense of the fact that the *Meteorologica* uses no teleological language: "and these things follow" explanations are not themselves teleological, and thus we should expect no teleological language to be used in describing them. Interestingly, neither should we expect to find teleological language used to describe the originative cycles from which they follow, for the teleology of the originative cycles is not established within the subfield of meteorology, but only in texts that take on the cosmic perspective from which these cycles are themselves viewed as *explananda*. We saw the teleology of organic generative processes ("man generates man") to be established by the cosmic perspective, not within the subfield of biology set out in *de Partibus Animalium*. Insofar as *de Partibus Animalium* uses teleological language, then, it is used in primary and secondary teleological explanations that have no counterpart in the subfield of meteorology.

Thus, establishing the details of imitation based on the activity of the prime mover seems to be out of place in both the *Meteorologica* and *de Partibus Animalium*, and we might even suggest that it is out of place in natural science more generally, despite some notable exceptions.[13] It seems that Aristotle's usual practice conforms to his proposal in the *Physics*: the "student of nature" needs to know the form "up to the point" of knowing that "man is begotten by man." Beyond that, knowledge of the prime mover as form is "the business of first philosophy to define" (II.2, 194b10–15).[14] In keeping with this suggestion, perhaps just as it is outside the purview of the natural science of *de Partibus Animalium* to consider form beyond the point of "man generates man," it is outside the purview of the natural science of meteorology to consider form beyond the point of the originative cycles.

In summary, I suggest that we should fill out the picture of natural scientific explanation as it applies in the fields of biology and meteorology as in Table 5.1.

It remains to be seen whether this schema tracks the actual explanations offered in the *Meteorologica* (the task of Section IV). All I have argued so far is that the disanalogy between inorganic and organic generative processes – the difference in their passive and active natures that takes root in their being partless wholes versus wholes of parts – should play out in the kinds of explanations we can expect to find in the biological and non-biological subfields of natural science. Whereas explanations in biology fill out the entire three-part schema of natural scientific explanation, explanations in meteorology are able to fit only the third "and these things follow" form of explanation. In the next section, I argue that the *Meteorologica* indeed lives up to these expectations.

13 For example, see *DC* II.1.

14 Through a close reading of *Physics* VIII, *de Motu*, and *GC* I–II, Andrea Falcon argues persuasively that Aristotle "enforc[es] the division of labor between physics and metaphysics" (forthcoming) "Aristotle's Integrated Study of Motion." In *Reading Aristotle*, eds. Ron Polansky and William Wians.

Table 5.1 Three forms of natural scientific explanation applied to biology and meteorology

Forms of Explanation	Text: *PA* 1.1, 640a25–b1	Biological Explanation	Biological *Explananda*	Meteorological Explanation	Meteorological *Explananda*
Primary Teleological Explanation	"For (γὰρ) man is generated from man; and thus (ὥστε) it is because the parent is such and such that the generation of the child is thus and so. Hence it would be best to say that, since (ἐπειδὴ) this is what it is to be a human being, on account of (διὰ) this it has these things; for it cannot be without these parts."	1. Man generates man. (Or its equivalent.) The ultimate *explanans*. 2. The (male) parent animal has a creative seed with the power to produce an offspring of the same form (implied by 1). 3. The offspring's parts come to be out of the best nutriment for the sake of realizing this form.	Vital and essential parts (e.g., the heart and liver are vital parts in blooded animals, and fins are essential parts in fish).	1. Originative cycles of (a) fire–earth and (b) air–water. The ultimate meteorological *explanantia*. 2. The elements have passive natures and are thereby such as to be transformed in due season by the sun in its circular course. 3. There is no analogue to primary teleological explanation because the elements have no parts to explain.	None
Secondary Teleological Explanation	"If one cannot say this, one should say the next best thing, i.e., either that in general it cannot be otherwise, or that at least it is good thus."	4.1. The formal nature uses inferior nutriment to make subsidiary parts that support the vital and essential parts produced according to 3, stated above. 4.2. The formal nature also co-opts the leftovers from processes 3 and 4.1, above, to make luxury parts for the sake of living well by providing further protection or support.	Subsidiary parts (e.g., kidneys, which support the functioning of the bladder). Luxury parts (e.g., nails, hair, hoofs, horns, beaks, and spurs of cocks).	4. There is no analogue to secondary teleological explanation because the elements have no parts to explain.	None
A Third, Non-teleological Explanation	"And these things follow" (ταῦτα δ᾽ ἕπεται).	5. Dregs remain as the by-products of the processes described in 3 and/or 4 above.	Parts or residues that do not perform any function (e.g., spleen, bile, and other residues of the gut and bladder).	5. There are by-products from the processes described in 1(a) and (b) above.	All meteorological phenomena besides the originative cycles themselves. (See Table 5.2 below for examples.)

Section IV "And these things follow" from the originative cycles

In this section, I argue that the *Meteorologica* takes the teleology of the originative cycles as given, and shows other meteorological phenomena to follow, as necessary by-products, from one or both of these cycles. So understood, Aristotle does indeed offer the "and this follows" third type of teleological explanation. Although I only have space to offer a sketch of my position, not a full detailed account, I chart the explanations of phenomena discussed in *Meteorologica* I–III according to my schema as in Table 5.2.

As my previous section showed from the cosmic perspective, the meteorological phenomena that compose the originative cycles are teleologically directed at imitating the prime mover. In this section, I aim to show that the other phenomena follow from these originative cycles according to the third, non-teleological, "and these things follow" mode of natural scientific explanation presented in *de Partibus Animalium* I.1. Although I will not be able to detail all of the examples presented above, I illustrate my point by examining two sets of paradigmatic cases.

Let us begin with Aristotle's explanation of winter rain and of summer rain. Winter rain composes an originative cycle with summer evaporation: when the sun recedes in the winter, air transforms into water, which moves downward, and when the sun approaches in the summer, water transforms into air, which moves upward. The sun ensures that these rectilinear movements occur on the heels of one another to form an imitative cycle. Summer rain, however, is not generated by the regular cooling due to the sun's annual winter recession, but rather by warmth due to the sun's proximity (perhaps also conjoined with the heat generated by the ignition of the fires as part of the earth–fire cycle [I.9 346b24]): "in the warmer seasons the cold is concentrated by the surrounding heat and causes the cloud to go over into water suddenly" (I.12, 348b6–8). Aristotle's explanation of summer rain relies on details of the originative water–air cycle: *regularly* in the summer the sun generates air, which rises and forms clouds, which *regularly* do not condense back into water until the sun recedes in the winter. However, this time something unusual happened: the heat of the summer sun surrounded the cloud and forced it to cool and condense into water out of season. Without reference to the originative cycle, Aristotle could not explain when and why this phenomenon took place, for the originative cycle explains the presence and location of the cloud and the heat in the first place. (Recall Aristotle's explanation of bile in Section I.) Summer rain, then, simply follows from the unusual interaction between factors that

Table 5.2 The phenomena of *Meteorologica* I–III indexed by originative cycles

Originative Cycle	Meteorological phenomena that are part of the originative cycle and thereby imitative and teleological	Meteorological phenomena that non-teleologically "follow from" one originative cycle	Meteorological phenomena that non-teleologically "follow from" the accidental confluence of multiple originative cycles
Earth–Dry Exhalation–Fire Cycle	Burning flames, torches, goats, shooting stars produced by combustion (I.5, 341b24–35), the Milky Way (I.8, 341b36).	1. Chasms, trenches, blood-red colors (I.6). 2. Shape of the flame (in the form of burning flames, torches, goats, and shooting stars produced by combustion [I.5] and the Milky Way [I.8]), which is determined by the length, breadth, and depth of the exhalation. 3. Oblique movement of the winds (II.4, 361a23–25).	1. Shooting stars and thunderbolts "produced not by combustion, but by projection under pressure" (I.5, 342a15–17, see also II.9 370a26–35). 2. Hurricanes (II.9 369a17–25 and III.1, 370b5–10). 3. The frequent appearance of comets within the zodiac circle (I.7, 344b28–345a10), due to unusually "excessive" winds (I.7, 344b19–32). 4. Dew formed in Pontus produced by recoil (ἀντιπερίστασις) (347a35–b7). 5. Hail and summer rain produced by recoil (ἀντιπερίστασις) (I.12). 6. Salinity of the sea (given its organic analogue as a residuum in II.2, 356a36–b3 and II.3, 358a17–27).
Water–Vaporous Exhalation–Air Cycle	Three kinds: 1. Diurnal: Nighttime dew and frost (I.10). 2. Annual: Winter rain and snow (I.11). 3. Great cycle: Rivers (I.14) and sea (I.14, 353a23–24, II.1–2).	1. Reflections of vapor, haloes, rainbows, and sundogs (III.2–6). 2. Rivers formed by trapped water that then forces its way out (ἀποβιαζομένου) 351a7–8).	7. Earthquakes (II.8, 366a4–5), which are a violent movement of the exhalation (σφοδρότατον 365b32, b33, σφοδρὸν 366a33, ἀποβιάζεσθαι 366b11) because the sun has less power (δύναται) over earthquake winds (II.8, 368b20). Earthquakes' organic analogues are pathological impulses in the body (II.8, 366b15–30). 8. Perhaps also minerals produced by the compression of exhalations enclosed in the earth (III.6, 378a15–b6).

usually contribute to an imitative cycle of elemental transformation. Since summer rain is generated as a result of this unusual confluence of factors, it has no generator to ensure its downward rectilinear movement will occur on the heels of the upward rectilinear movement of air. Thus, summer rain is a non-imitative rectilinear dead-end, what Aristotle calls "violent rain" (ὕδατα λαβρότερα, 348b11, 348b23 or ῥαγδαῖα, 349a7), thereby contrasting it with its teleologically imitative counterpart, winter rain.[15]

Just as Aristotle sharply contrasts the production of winter rain and summer rain, he distinguishes shooting stars produced as part of the originative cycle from shooting stars produced by ejection occasioned by the accidental confluence of the water–air with the earth–fire cycles. Regularly, as part of the originative earth–fire cycle, the superlunary bodies cause shooting stars to be "formed in the upper region . . . due to the combustion of the [dry] exhalation" (342a16). However, sometimes shooting stars result ("these things follow") as an accidental by-product of the confluence of the water–air cycle with the earth–fire cycle:

> When it takes place at a lower level it is due to the ejection of the exhalation by the condensing and cooling of the moister exhalation; for this latter as it condenses and inclines downwards contracts, and thrusts out the hot element and causes it to be thrown downwards. (342a16–21)[16]

Once again, Aristotle explains this occurrence by way of a "these things follow" explanation that traces the phenomena back to the originative cycles: as we have seen above, the cooling and condensation of the moist exhalation is a regular part of the water–air cycle, but in this instance the upward movement of earth's dry exhalation, which usually gets combusted and transformed into fire by the superlunary motions, gets caught up in the water–air cycle, thereby taking "a course more like that of a projectile than of a fire" (I.5, 342a2–3). This unnatural direction of fire's movement under ejection, which has no immediate superlunary origin, stands in sharp contrast to the natural upward movement of fire as shooting stars produced by combustion caused by the superlunary motions. Just like summer rain, such accidentally produced shooting stars are non-imitative rectilinear dead-ends.

15 As previous papers of mine argue in depth, the best reading of *Phys.* II.8 shows Aristotle committed to winter rain and summer heatwaves being teleological (on their own and independent of biological processes) and to summer rain and winter heatwaves being accidental (Scharle 2008; "Man from Man, not Bed from Bed: Nature, Art and Chance in *Physics* II." In *Aristotle's Physics: A Critical Guide*, Cambridge: Cambridge University Press, 2015). My reading is also supported by the Arab commentary tradition on the *Meteorologica.* See Lettinck (1999, 97–119).

16 See also II.9, 370a26–35; and hurricanes at II.9, 369a17–25 and III.1, 370b5–10.

This section illustrates that Aristotle does indeed offer the "and this follows" third type of teleological explanation in the *Meteorologica*: he takes the teleology of the originative cycles as given, and shows other meteorological phenomena to follow, as necessary by-products, from one or both of these cycles. Although I do not have the space to demonstrate this for all of phenomena catalogued above, it is a significant finding that at least some can be traced back to these two cycles. This shows that teleology can indeed play a role in meteorological explanation.

Section V Conclusion

This chapter suggests a unified theory of Aristotle's treatises. This theory illuminates subtle patterns in the focus and explanatory approach to each subject and explains the lack of teleological language in *Meteorologica*. Scholars have disagreed about the philosophical issue at stake in Aristotle's introduction of formal and final causation on the organic level: one side claims Aristotle's reason was to explain the organic phenomena his opponents tried but failed to explain via elemental material and efficient causation, and the other side claims his reason was to refute a theory on which organic phenomena would fail to be *explananda*.[17] Despite this disagreement, scholars on both sides have been united in assuming that Aristotle's defense of natural teleology solely targets his predecessors' conception of organic phenomena: scholars have believed that while Aristotle agrees with his predecessors that the four elements operate solely by efficient and material causes, he held that the generation of living things requires, in addition, formal and final causes.[18]

My new reading of the *Meteorologica* questions this common assumption. If I am right, my reading of the *Meteorologica* supports Sarah Broadie's suggestion that "natural change is ontologically and conceptually presupposed by change of all other types."[19] In the *Meteorologica* we have traced meteorological phenomena back to one or both of the natural, teleological originative cycles. We might go a step further, then, to suggest that *teleological* change – even at the level of the elements – is ontologically and

[17] Anti-reductivist interpreters include Gotthelf (1987; 1997), Charlton (1992), Waterlow (Broadie) (1982), Cohen (1989), Cooper (1982), Charles (1991), Bradie and Miller (1984), and Balme (1987). See Gotthelf (1997) for a useful taxonomy of these views. Anti-eliminativist interpreters include Sauvé Meyer (1992), Irwin (1988), and Johnson (2005).

[18] The following commentators explicitly accept the claim: Cooper (1982, 202), Bradie and Miller (1984, 141), Charles (1988), Lewis (1988, 55), Nussbaum (1978, 55), and Gill (1997, 147). See also Waterlow (Broadie) (1982, 92).

[19] Waterlow (Broadie) (1982, 38).

conceptually presupposed by change of all other types. This would show that Aristotle's dispute with his materialist predecessors runs much deeper than scholars have thought.

My interpretation of the *Meteorologica* explains why some scholars might have missed the crucial role teleology plays on the elemental level[20] and why this role is so easy to miss. The *Meteorologica* describes only two teleological cycles by contrast with the many, many teleological processes found in the biological treatises. These two proportions may perhaps be explained by the household analogy of *Met.* XII.10, which suggests that the heavenly bodies "have least license to act as they chance to, but all or most of what they do is arranged (τέτακται)," while the sublunary elements "can do a little towards what is communal, but act mostly as they chance to" (1075a19–22).[21] Living things are able to co-opt and refashion by-products into teleologically directed subsidiary and luxury parts – something that the elements are denied, possessing solely passive natures. Without this ability, the elements mostly operate as they chance to, which is just what the household analogy suggests. I have argued that what they mostly chance to do – as by-products – follows from the little they do toward what is communal – as participants in originative cycles – and this fact is significant in showing the priority Aristotle affords teleology even at the level of the elements.

20 Scholars often think that Aristotle's natural teleology applies exclusively to biological things (plants and animals) and that the elements (earth, air, fire, and water) either are not teleological, or are teleological only insofar as they play a role in biological processes. Gotthelf and Lennox (1987) and Lennox (2001a) are paradigmatic examples of the former approach, while Sedley (1991) is of the latter. Notable exceptions include Johnson (2005), Quarantotto (2005), and Leunissen (2010).

21 Trans. Sedley (2000, 328). For a fuller discussion of this passage, see Scharle (2008).

CHAPTER 6

Aristotle on the cosmological significance of biological generation

Devin Henry

Introduction: two perspectives

David Sedley is perhaps the most well-known defender of the view that attributes to Aristotle a thoroughgoing and comprehensive global teleology in which the entire contents of the natural world – from living things and their parts, to the elements, to the seasons themselves – are so arranged that their mutual interactions contribute to the overall good of the universe and ultimately (in the case of sublunary phenomena) to the good of man.[1] It follows from this that Aristotle's natural teleology cannot be fully grasped except from this comprehensive global teleological perspective.[2] In opposition to this, many modern scholars defend what I will call the 'organism-centred' view of Aristotle's natural teleology. This reading treats the individual organism as the final end for the sake of which all its features exist, so that all final causation in biology can be understood exclusively from the perspective of the individual organism itself. One of the fundamental suppositions of Aristotle's natural science, for example, is that nature does nothing in vain but always what is best given the range of

This chapter was read to the Princeton Classical Philosophy Colloquium (December 7, 2013). I am grateful to all the participants for their questions and comments and especially to my commentator Sean Kelsey, as well as James Lennox and Ben Morison for our correspondences.

1 In this chapter I will be concerned only with the use of teleology as it applies to living nature. It is a matter of some controversy whether or not Aristotle thinks natural teleology applies to non-living nature in any robust sense. For example, one could accept that the motion of the elements towards their proper place is directional and yet deny that Aristotle thinks that motion is to be explained by reference to the contribution it makes to an element's good, which I take to be a central feature of Aristotle's teleology. For an alternative view see Johnson (2005, Ch. 5), Scharle (this volume), and Falcon and Leunissen (this volume). All translations are my own except where indicated.

2 See Sedley (1991: e.g. 180; 2007, 167–204; 2010, 18–29). Sedley (2010, 24) argues that for Aristotle 'biological teleology is derivative from a prior cosmic teleology'. Like Plato, he held that 'the world's goodness is the primary *explanandum*, and that the same world's comprehensive stocking with organic species, each elaborately engineered to suit its specific function, is in turn explained by its contribution to that goal'. This view is also defended by Kahn (1985) and Furley (1985).

possibilities. And Aristotle is explicit that 'what is best' is understood in relation to the organism's *own* survival and well-being (*IA* II, 704b12–18).[3] This conforms to Aristotle's practice in the biological works: there is no attempt on Aristotle's part to show how any of the adaptations possessed by living things contribute to the good of the universe as a whole. Instead Aristotle assumes that each individual has the features it does because, and only because, those features are good for that individual.[4]

As a general interpretation of Aristotle's natural teleology I think the organism-centred view has it right. For the most part Aristotle treats the survival and well-being of each individual organism as the final end for the sake of which its features ultimately exist. And while those features might also benefit things outside the individual organism, those benefits are not part of the causal story that explains why they are present in the first place. At the same time I think there is value in considering an interpretation like Sedley's insofar as it forces us to take seriously the connections between Aristotle's natural teleology and his broader cosmology. In my view the collective reaction to Sedley's reading has had the effect of swinging the pendulum too far in the opposite direction. As a consequence, scholars have tended to ignore, or at least downplay, the cosmological features of Aristotle's biology. To be sure I think Sedley overestimates his case by attributing to Aristotle a thoroughgoing global interactionist teleology. Indeed, as far as I can tell, there is only one exception to the organism-centred view (discussed below). But this is hardly enough to vindicate the Cosmic Teleology reading. For the explanation in question is not part of any systematic attempt by Aristotle to adopt a more global teleological perspective of the sort defended by Plato (e.g. *Phaedo* 98b1–4, *Laws* 903b5–d1). However, I think this one passage is enough to show that Aristotle held a

3 I discuss this principle in Henry (2013).

4 For a classic statement of this view see Balme (1972, 96–97). It is defended by Bodnar (2005), Lennox (2001a, e.g. 182–84; 2001b, 341), Judson (2005, 348, 359–60), Gotthelf (2012, e.g. 8–9 n. 13), and Leunissen (2010, e.g. 41: 'cosmic teleology is ultimately grounded in natural teleology' and 'natural teleology in its primary form is limited to individual natures acting as efficient causes for the sake of something'). *Phys.* II 7, 198b5–9 is among the main passages cited in defence of the view that, for Aristotle, each organism's nature aims at its *own* good. Lennox (2001b, 341) calls this Aristotle's 'basic teleological axiom'. Not everyone who rejects cosmic teleology accepts the organism-centred view. Johnson (2005), for example, argues that each particular substance is the final end for the sake of which all its features exist, so that all final causation can be understood exclusively from the perspective of the particular substance itself (e.g. p. 278: 'the good which teleological explanations make reference to is specific to the natural kind being explained'). It just turns out that, on his reading, organisms are not the only natural substances subject to teleological explanation. Aristotle extends his doctrine to the elements and their natural motions. See also Scharle (2008).

weaker version of the organism-centred view than has traditionally been ascribed to him by opponents of Sedley.[5]

The exception I have in mind comes at the beginning of *Generation of Animals* (*GA*) II where Aristotle is attempting to give an account of the final causes of sexual differentiation. I claim that Aristotle's argument here is a specific application of a more general argument he gives in *Generation and Corruption* (*GC*) II 10, which seeks to explain why coming to be and passing away occur continuously and without fail by placing them in the context of his broader cosmology. As we shall see, what is at stake with this passage is exactly how to understand the teleological significance of animal generation. If I am right, then what Aristotle is arguing in *GA* II 1 is that continuous animal generation exists, not (or not only) because of the contribution it makes to the individual's own good but because of the contribution it makes to the good of the universe as a whole. Before turning to that, let me begin by drawing attention to another passage to give a sense of what I mean by 'the cosmological aspects' of Aristotle's biology.

In *GC* II 10 Aristotle describes how the sun's annual motion around the earth causes generation and destruction in the sublunary world as it approaches and retreats. Part of that story includes an account of how this motion influences the life-cycles of living things:

> This explains why it is not the primary motion that causes generation and destruction, but the motion along the inclined circle [sc. the sun's annual motion in the ecliptic] . . . the approaching and retreating of the moving body are caused by the inclination. For the consequence of the inclination is that the body becomes alternately remote and near . . . Therefore, if it generates by approaching and by its proximity, it – the very same body – destroys by retreating and becoming remote . . . Hence, too, the times (i.e. life-cycles) of the several kinds of things have a number by which they are distinguished. For there is an order for all things, and every time (i.e. every life-cycle) is measured by a period. Not all of them, however, are measured by the same period, but some by a smaller and others by a greater one. For some of them the period that is their measure is a year, while for others it is longer and others shorter than this. (*GC* II 10, 336a33–b16 Joachim translation with modification; cf. *GC* II 11, 338b3–5)

In *GA* IV 10 Aristotle explains how this works in the case of animal generation (777b16–778a9). There he tells us how the lives of organisms 'strive

[5] To be clear, what I am advocating in this chapter is a weaker version of the organism-centred view, *not* a more limited version of the Cosmic Teleology reading. I take this one very limited case of global teleology to be an exception to Aristotle's otherwise thoroughgoing organism-centred view where each species of living thing has the adaptations it does because *and only because* they contribute to *its own* survival and well-being.

to be measured according to natural cycles' (i.e. days, months, and years), which are determined by the motions of the sun and moon:

> For the month is a common cycle between these two [sc. the sun and moon], and the moon is a principle owing to its communal relationship with the sun and its participation in its light; for it becomes just like a lesser sun. For this reason the moon contributes to all processes of development and completion. For the heatings and coolings, in a certain proportion, produce generation followed by destruction. And the limits of these (both their starting-points and terminating-points) are contained by the motions of the stars. For just as we observe a sea and everything that has a fluid nature being stabilised and changed according to the motion and stillness of the winds, and the air and the winds according to the cycles of the sun and moon, so too both the things that grow from these and other sorts of things necessarily follow upon them. For it is rational for the cycles of the less important things to follow upon those of the more important things. For even wind has a kind of life, a generation and destruction. And there are probably certain other sources of the revolutions of these stars. Therefore, nature seeks to count out the processes of development and completion by the numbers of these. But it is not precise owing to the indeterminacy of the matter and because of the many principles that impede the natural processes of generation and destruction and frequently cause things that happen contrary to nature.

The explanation here has two parts. The *efficient causal* explanation attempts to identify the mechanism by which the stages of an animal's life-cycle are ultimately controlled by the motions of the sun and moon. Their motions cause changes of the seasons, bringing with them changes in temperature and light, which in turn trigger events within the organism associated with the phases of its life (e.g. as the days get shorter there is a decrease in light and temperature that triggers the hibernation instinct in certain animals). But Aristotle also thinks the phenomenon in question can be given a *teleological explanation*. Aristotle is not suggesting that the cosmic cycles occur for the sake of regulating the phases of an animal's life: they do not occur *in order that* animals should have something by which to measure their life-cycles. Instead he is saying that the formal natures of animals *use* the motions of the sun and moon and the changes that follow upon them for that purpose. And being 'used for the sake of something' is a kind of teleological relation.[6]

I draw attention to this passage in order to illustrate how Aristotle thinks that the full significance of certain aspects of animal generation

[6] See Leunissen (2010, e.g. 81–99).

cannot be grasped without reference to cosmology. For the explanation on offer depends, at least in part, on understanding Aristotle's broader theory about the causal dependence of changes in the sublunary world on the eternal motions of the heavens that he articulates in *GC* II 10. Of course in this particular example the appeal to teleology does not involve relating the feature in question to some broader cosmic good, and so this observation is entirely compatible with the organism-centred interpretation. Thus, while the *GA* IV 10 passage illustrates *the cosmological significance* of animal generation, it does so without depending on *a global teleological perspective* of the sort articulated by Sedley. But the same cannot be said for the *GA* II 1 passage that forms the focus of this chapter. In that passage Aristotle adopts the global teleological perspective in order to explain the feature in question. Or so I shall argue.

It is worth stressing at the outset that the arguments in this chapter should in no way be taken as a defence of Cosmic Teleology. I think that reading fails as an overall interpretation of Aristotle. Instead my aim is to advocate for a slightly *weaker* version of the organism-centred view than opponents of Sedley have traditionally defended. While it is true that, for the most part, Aristotle treats the organism's own good as the final end for the sake of which its adaptations ultimately exist, I want to stop short of claiming that *all* final causation in Aristotle's biology can be understood *exclusively* from the organism's perspective. Instead I shall attempt to show that there is at least one feature – the process of biological generation itself – that cannot be understood from the perspective of the individual organism alone but must be explained (at least in part) by the contribution it makes to the good of the universe as a whole. Or so Aristotle seems to claim. At the same time, this is hardly enough to vindicate the Cosmic Teleology reading. For (as mentioned) it is not part of any systematic attempt on Aristotle's part to adopt a more global perspective of the sort we find in Plato.

Why are there sexes? *Generation of Animals* II 1

One of the major puzzles driving Aristotle's *Generation of Animals* is why there are sexes. Of particular importance to the case of animal generation is why animal species are divided into separate sexes.[7] *GA* II 1 opens by introducing two modes of explanation for this, one that refers to the embryological mechanisms responsible for sexual differentiation (efficient

[7] Aristotle thinks that plants have male and female principles but that these are present together in the same individual. The *aporia* of *GA* II 1 is why they are separated in the case of animals.

causation), and one that refers to the advantage that comes from having separate sexes (final causation).[8] The opening of Book II provides the teleological explanation, while the efficient cause explanation is postponed until Book IV.

The teleological explanation proceeds in two steps. First we get an argument for why the sexes exist at all (731b24–732a3). Aristotle then goes on to show why animals are separated into sexes, that is, why some individuals are male and others female (732a3–10). In what follows I shall focus exclusively on the first argument. The key text reads as follows:

> [A] Of the things that are, some are eternal and divine, while others are capable of being and not-being, and the beautiful and divine is always a *per se* cause (*aition aei kata tên hautou phusin*) of what is better in things that admit it. And what is not eternal is capable of being <and not being> and admits of the better and worse. And soul is better than body, and what is ensouled is better than what is not owing to the possession of soul, and being is better than not being, and living is better than non-living. These are the causes owing to which the generation of animals exists. [B] For, since it is not possible for the nature of this kind of thing to be eternal, what comes into being is eternal in the only way possible for it. While it is not possible to be eternal *in number* (for the substance of existing things is in the particular; if it was of this sort, then it would be eternal), it is possible to be so *in form*. For this reason there is always a continuous generation[9] of humans, animals, and plants. [C] Since the principle of these is the male and female, it is for the sake of generation that the sexes are present in those that possess them. (*GA* II 1, 731b24–732a3)

In [A] Aristotle divides existing things into those that are everlasting and enjoy eternal being (e.g. god and the heavenly bodies) and those that are only capable of temporary being (the elements, animals, and plants). The latter alone are subject to generation and destruction since they contain matter, which is the seat of a thing's capacity to be and not be (*GC* II 9, 335a33–34, *Metaph.* VII 7, 1032a20–22). Aristotle next invokes a series of axiological principles (e.g. being is better than non-being, living is better than non-living), which are said to provide 'the causes owing to which the generation of animals exists'. [B] then expands on this point. While no sublunary organisms are capable of existing indefinitely (each one is composed of matter and so eventually decays), they can become eternal in form by engaging in reproduction. The upshot of this, Aristotle says,

8 The formal and material causes of the sexes themselves are found in *GA* I 2.

9 I follow Lennox in taking *genos* here to have the sense given at *Metaph.* V 28, 1034a29–30: 'the continuous generation of things having the same form' (in modern biological terms, an ancestor–descendent lineage).

is that, while there cannot be eternal *organisms* in the sublunary realm, there can be eternal *lineages* ('For this reason there is always a continuous generation of humans, animals, and plants'). [C] This, in turn, provides the final cause for the existence of males and females: male and female are 'principles' of generation (*GA* I 2, 716a5–8) and so are conditionally necessary for reproduction.[10]

Traditionally, this passage has been read in the context of a similar argument from *DA* II 4. There Aristotle argues that the capacity for reproduction exists for the sake of the individual organism insofar as it provides her with a means for partaking in the divine:

> For this is the most natural among the functions for living things . . . namely, to produce another like itself, an animal producing an animal, and a plant producing a plant, in order that it may partake in the everlasting and divine insofar as it is possible. For every living thing desires this and performs all of its natural activities for the sake of it . . . Since, then, they cannot share in the immortal and divine by continuous existence, for no perishable thing can remain numerically one and the same, they share in these in the only way they can, some more and some less. What persists is not the individual itself but something like it, not numerically one but one in form. (*DA* 415a26–b7)

According to this passage, the significance of reproduction is that it grants the individual access to 'the immortal and divine', where what persists is not numerically the same individual but something that is similar to it in form. Most scholars take it for granted that our passage from *GA* II 1 simply reiterates this point. When read in that context, *GA* II 1 is understood, again, as arguing that reproduction exists for the sake of the individual insofar as it provides a means for achieving the only kind of immortality open to her.[11] On this reading the continuity of the species is not the end for the sake of which reproduction exists but a mere consequence of the fact that, when every individual strives after its own good, what you get is a continuous generation of things of the same form. We can call this the organism-centred reading of *GA* II 1, since it treats the good of the individual as the final cause for the sake of which animal generation exists and eternal lineages as a mere by-product of this.

10 This reading of *GA* II 1 agrees closely with Lennox (1985, 71), though we disagree over the good for whose sake continuous generation takes place (see below).

11 Kullman (1985, 172) writes of the *GA* II 1 passage: 'The idea is, once again, that the circle of procreation is the way in which animals, inasmuch as they are sexual, partake of the everlasting and the divine, which in this respect functions as the final cause.' See also Balme (1972, 96; 1987, 279–80), Lennox (1985, e.g. 67, 73, 92 n. 9), Leunissen (2010, 41), and Gotthelf (2012, 9 n. 13, 58, 71).

In contrast to this, I want to argue that *GA* II 1 is not just another version of the *DA* II 4 argument. Instead it should be read as a specific application of a more general argument that Aristotle makes in *GC* II 10, which seeks to explain why coming to be and passing away occur continuously and without fail. What is at stake between these two readings, we shall see, is exactly how to understand the teleological significance of generation according to the *Generation of Animals*. If I am right, then what Aristotle is arguing in *GA* II 1 is that continuous generation exists not (or not only) because of the contribution it makes to the individual's own good but because of the contribution it makes to the good of the universe as a whole.[12] In this way *GA* II 1 not only highlights the cosmological significance of animal generation; it also attempts to explain that phenomenon from a global teleological perspective. To see this, we need to examine *GC* II 10 in more detail.

The continuity of generation: *Generation of Animals* II 10

In *GC* II 10 Aristotle asks, what is the cause responsible for the fact that generation and corruption go on continuously and never fail? He has already provided the material cause of the phenomenon back in *GC* I 3.[13] *GC* II 10 offers the efficient and final causes.

Efficient causal explanation **(336a15–336a7).** In the *Physics* Aristotle established that locomotion is the primary form of change and that this is what causes generation and destruction. In *GC* II 10 he argues, more specifically, that this is the sun's annual movement in the ecliptic as it approaches and retreats from the earth's surface (cf. *Meteor.* 346b20–22). This, Aristotle says, can be supported by appealing to the observable facts: 'for we see (*horômen*) that generation occurs as the sun approaches and things decay as it retreats' (336b16–24). Since the sun's motion is eternal, it follows that generation and destruction must themselves occur continuously and without fail.

[12] For a similar reading see Kahn (1985, 194–96) and Cooper (1982). Cooper recognises that *GA* II 1 treats the continued existence of living things itself as an important good and not just the existence of the individual organism, but he does not trace the argument to *GC* II 10. By contrast, while Gotthelf (2012, 58) sees the connection between *GA* II 1 and *GC* II 10, he continues to treat the former as if it were simply a version of the *DA* II 4 argument. On his reading, the lesson of *GA* II 1 is that reproduction is aimed at *self*-preservation, while the existence of eternal lineages is merely a consequence of that. Finally, Kahn's interpretation, though similar to mine, embraces a cosmic teleology of far greater magnitude than anything I am willing to endorse.

[13] See 319a18–22.

Teleological explanation **(336b25–337a7).** Aristotle next gives us a teleological explanation for the continuity of generation and destruction in the sublunary realm, which appeals to the notion of what is best. The main text of the argument runs as follows:

> As we have said, coming to be and passing away will be everlasting and continuous and will never fail owing to the reason stated. This conclusion also follows rationally (*eulogôs*). For, since we say that nature always strives after what is better, and that being is better than not being . . . but not all things can possess <eternal existence>[14] because they are too far removed from the first principle, it follows that the god adopted the remaining alternative and filled up (*suneplêrôse*) the universe by making generation perpetual. For that way existence would be most connected (*malista suneiroito*) because the fact that coming to be should itself come to be perpetually is the closest approximation it has to eternal being. (*GC* II 10, 336b25–35 translated after Joachim)

I'll set to one side the fact that this passage seems to rely on the unAristotelian idea that the universe was created by a providential god who deliberated about what is best for the world. It is almost certainly metaphorical.[15] One suggestion is that Aristotle is simply invoking the popular idea of a creator god as a heuristic device. Reasoning *as if* the universe were designed by a providential god aiming at some optimal state of affairs is a useful heuristic device because it helps us identify the final cause of continuous generation. Aristotle, like Plato, takes it as a basic fact, not explicable in terms of anything more fundamental, that the universe is an ordered system and that this order is good. It follows that everything that contributes to this order is itself good. But that does not commit Aristotle to the further claim that there must have been some Divine Craftsman who organised the parts of the universe in order to make it so.[16]

The conclusion of the above argument is fairly clear. Continuous generation exists in the sublunary world because it is the best possible way to maximise the level of being in the universe as a whole given the constraints on sublunary being. Aristotle deduces this conclusion from two universal principles, both of which are taken as basic:

[14] The point of the argument seems to be, not that some things are incapable of existing at all (e.g. square triangles), but that some things are not capable of *eternal* existence. This is shared premise of both the *GA* II 1 and *DA* II 4 arguments.

[15] Even Sedley (2007: 168 n. 4) dismisses the language of god creating the world as figurative.

[16] I owe this suggestion to Sean Coughlin. On the use of teleological principles as heuristic devices see Leunissen (2010, 119–35).

(i) Nature always strives after what is better.
(ii) Being is better than non-being.

The first of these is explicitly identified as a first principle of natural science in *IA* II, which states that 'nature does nothing in vain but always what is best for the substance from among the possibilities concerning each kind of animal' (704b14–15).[17] This assumption is supposed to license a certain pattern of explanation that appeals to the concept of what is best (what we call 'optimality' reasoning). According to this pattern of reasoning, we explain why something is the way it is by showing how being in that state is *the best possible* way for it to be. As Aristotle puts it, 'if it is better this way, then it is that way and being in that state is in accordance with nature' (*IA* 704b17–18).[18]

The teleological argument in *GC* II 10 exhibits this pattern of reasoning. Since being is better than non-being, the ideal universe would be one that contained only eternal substances that never passed out of existence. That way the universe would be completely filled up with being with no traces of non-being. But since the actual universe contains a sublunary realm of material composites, and since all material composites must come to be and pass away (*GC* II 9, 335a32–b5), this state of affairs is not possible. The alternative, Aristotle reasons, is to have an everlasting cycle of generation that guarantees an endless supply of generated beings: 'That way', he says, 'existence would be most connected, because the fact that coming to be should itself come to be perpetually is the closest approximation it has to eternal being.' If the generation of sublunary beings were intermittent rather than continuous, so that there were periods in the earth's history when no sublunary beings existed, that would decrease the overall amount of being in the world. The world would contain existential gaps, as it were.[19] Perpetual generation avoids this by linking individuals together in a continuous (gap-less) chain, so that, although each individual must pass out of existence, there will always be some sublunary beings in existence.[20]

[17] Aristotle treats this as an empirical truth based on observations about the way living things behave (e.g. *de Juv.* 469a28–30).

[18] I see this kind of optimality reasoning, which explains why *S* is *P* by showing that *P* is 'the best way' for *S* to be, as a sub-species of teleological explanation even though it does not typically involve the same 'for the sake of' language characteristic of more familiar kinds of teleological explanation. Compare *Phaedo* 97b8–98a2 where Socrates appeals to what is best as a cause (*aitia*) of things in nature. For a discussion of this type of explanation in Plato and Aristotle see Henry (2013). For another example of where Aristotle takes the appeal to what is best as explanatory (i.e. as part of the *aitia* of a thing) see *IA* 708a9–20.

[19] I owe this way of expressing the point to Chris Frey.

[20] This way of reading the argument does not commit Aristotle to the claim that the universe must somehow be benefited by continuous generation (cf. Johansen, this volume). All it commits Aristotle

What I want to suggest is that the *GA* argument means to apply this general argument to the specific case of biological generation in order to explain why the sexes exist. Read in this context, the *GA* argument can be reconstructed as follows:

(1) Nature always strives after what is better given the range of possibilities.
(2) Being is better than non-being, living (which is being for an organism) is better than not living, and so forth.
(3) No sublunary organism is capable of eternal life; since they contain matter, such organisms must come to be and pass away.
(4) Therefore, the god adopted the only remaining alternative and filled up the living world completely by making biological generation perpetual.
(5) This is why there is continuous generation of humans, animals, and plants.

The core idea in this argument is that, since there cannot be any eternal organisms in the sublunary world insofar as they contain matter, the next best thing is to make their generation continuous. For having eternal lineages is the best possible way to maximise the amount of life in the world given the constraints imposed on living things by their material natures. Eternal lineages would ensure that biological existence is 'most connected' (*malista suneiroito*) by leaving no existential gaps in the living world. This, in turn, provides the final cause for the existence of sexes: males and females are present for the sake of this end, insofar as they constitute the first principles of generation.[21]

To close this section, let me offer two reasons for thinking the *GA* passage is an extension of the *GC* II 10 argument and not a version of the argument from *DA* II 4. First, the *GA* II 1 argument resembles the *GC* II 10 passage more than it does *DA* II 4. The main reason scholars have tended to read the *GA* and *DA* passages together is that text [B] is a shared premise between the two arguments. But we can now see that [B] is also a premise in the *GC* II 10 argument. And so this by itself cannot help to decide between the two readings. By contrast, the *GC* and *GA* arguments both draw on the same axiological principles expressed in [A], which do not figure into the *DA* argument. While it may be conceded that these same principles are in

to is the claim that (a) there is some best way for the universe to be and (b) continuous generation exists *because* it contributes to that good state.

[21] See Lennox (1985, 72–73). Because Lennox reads *GA* II 1 in the context of *DA* II 4, however, he thinks continuous generation occurs because it procures the best possible state of affairs for the individual organism (1985, 72).

the background of the *DA* argument, the fact that Aristotle makes explicit use of them in the *GA* passage creates a strong presumption in favour of my reading. For it suggests that Aristotle had *GC* II 10 in mind when he composed that passage (cf. *GC* 336b28–9).

Second, and more importantly, the *DA* and *GA* arguments are not focused on the same aspects of generation. The *DA* argument is concerned with the individual's need to reproduce, which Aristotle thinks is an expression of its desire for immortality. The *GA* argument, by contrast, makes no mention of this. There is no reference anywhere in the text to the idea that individuals engage in reproduction in order that they may 'partake in the everlasting and divine insofar as it is possible' (the key premise of the *DA* argument). Instead, the *GA* passage is focused squarely on the continuity of generation itself and, ultimately, on why sexes exist.[22] Neither of these is at issue in *DA* II 4. It is true that the *DA* argument presupposes the continuity of generation. The fact that Aristotle thinks individuals are members of eternal species is what grounds his claim that formal replication provides a means for individuals to partake in the immortal and divine. If he thought it were possible that species might go extinct, then reproduction could hardly be said to offer a way for individuals to share in everlasting existence. But the point of the *DA* argument is not to explain why species continue indefinitely into the future but why individual organisms engage in reproduction. And the answer to that question is, so that they can partake in the immortal and divine. As I read the *GA* passage, texts [A] and [B] constitute a single argument whose conclusion is given at 731b35–6: 'This is why (*dio*) there is always a *genos* of humans, animals, and plants.' This way of understanding the passage makes it clear that the *explanandum* of the argument all along was the fact that generation continues eternally and without fail. And that is exactly what the *GC* II 10 argument was meant to explain at a more general level.

In this chapter I have offered two interpretations of the *GA* II 1 argument, which have quite different implications for how he understood the teleological significance of biological generation. According to the organism-centred interpretation, the *GA* passage is merely echoing the point of the *DA* argument: males and females exist for the sake of reproduction, and reproduction exists for the sake of the individual. On this reading, generation is explained by the contribution it makes to the *individual's own* good

22 See below. It is fairly standard to read the *GA* II 1 argument as focused on the continuity of generation itself (e.g. Cooper 1982). Even Gotthelf, who defends the organism-centred reading, takes the passage to be 'explaining the unendingness of sexual generation' (2012, 58). This is especially true if Lennox is right that *genos* at 732a1 has the sense given at *Metaph.* V 28, 1024a29–30, which refers to 'a continuous generation of things having the same form' (i.e. a reproductive lineage).

(it provides the best possible means for individuals to achieve a share in the immortal and divine). The fact that there is a continuous succession of living things is simply a consequence of individuals trying to maximise their good in the only way they can but is not the end for the sake of which reproduction exists. By contrast, I take the argument in *GA* II 1 to be focused on the existence of eternal lineages themselves. On this reading, eternal lineages are treated, not as a mere side-effect of individuals striving after their own personal good, but as the best possible means of 'filling up' the universe with being, given what is possible in the realm of living things. In this context, continuous generation is explained by the contribution it makes to the good of the universe as a whole (following the *GC* II 10 argument). The sexes exist because they are necessary for this. This gives the sexes a more cosmological significance.[23] Does this mean that Aristotle endorsed a version of Cosmic Teleology as Sedley claims? In the next section I would like to offer some reasons why we should resist this move.

Cosmic teleology revisited

As noted *GA* II 1 begins by introducing two modes of explanation for sexual difference, an efficient-causal story that gives the embryological mechanisms responsible for differentiating individuals into males and females, and a teleological account that refers sexual difference to 'a higher principle' (*anôthen*). On the reading defended here *anôthen* is meant to connect the cause of sexual difference to Aristotle's broader cosmology.[24] As we have seen, one of the central puzzles of Aristotle's cosmology is why there is continuous generation in the sublunary world. He answers this question in *GC* II 10 partly by appealing to the causal influence of the sun's eternal rotation along the ecliptic and partly by invoking principles of optimisation that show how continuous generation is the best possible way of maximising the amount of being in the sublunary world. I have argued that *GA* II 1 is an attempt to apply this same pattern of reasoning to

[23] There is at least one other place where this is suggested. In *GA* IV 3 Aristotle tells us that females exist, in part, because they are necessary to preserve the continuity of each sexually differentiated animal species (767b8–10: *sôzesthai to genos tôn kechôrismenôn kata to thêlu kai to arren*). However we translate *genos* here, Aristotle is clearly saying that sexes exist for the sake of preserving something more inclusive than the individual.

[24] Compare *GA* I 2, 716a13–16. Balme's epistemological reading of *anôthen* as a reference to a 'more universal' principle cannot be ruled out (cf. *APo.* 97a33). The more universal principles in question, I take it, would be the various axiological claims that Aristotle makes about the relative value of soul, life, and being. But this is consistent with my broader reading of the passage insofar as the cosmological argument in *GC* II 10 also derives its conclusion from universal principles of the same sort.

explain why there is a continuous generation of living things. Since no individual organism is capable of eternal existence, the closest approximation to eternal being in the living world is to make generation perpetual. That way life would be 'most connected' (*malista suneiroito*), as Aristotle puts it in *GC* II 10. This, in turn, provides the final cause for the existence of the sexes insofar as males and females are first principles of generation. On this reading, continuous biological generation exists, not (or not only) because it contributes to the good of the individual, but because it contributes to the good of the cosmos as a whole.[25]

Having said that, I think it would go too far to treat this as evidence for Sedley's Cosmic Teleology reading.[26] On Sedley's reading, Aristotle thinks the universe is an organised whole endowed with a nature of its own. This 'cosmic nature', Sedley argues, is something over-and-above the natures of its individual parts (animals, plants, etc.). Sedley's main evidence for this reading comes from the controversial passage at the start of *Metaph.* XII 10. There Aristotle considers the way in which 'the nature of the whole (ἡ τοῦ ὅλου φύσις) contains the good and the best, whether as something separated and by itself, or as its arrangement' (see *Metaph.* XII 10, 1075a11–25). Sedley takes the reference to 'the nature of the whole' to pick out a cosmic nature that belongs to the universe *as a whole* and embodies its good. This cosmic nature, Sedley argues, is prior (and thus irreducible) to the natures of the individual organisms, since the latter are parts of the former.

I am not unsympathetic to Sedley's claim that Aristotle recognised more inclusive individuals above the level of particular organisms and that these more inclusive individuals might have natures of their own. In the *Politics* Aristotle has no trouble seeing the *polis* as a natural whole that is ontologically prior to the citizens, which are its parts. So it may not be much of a stretch to imagine him treating the cosmos itself as a complex individual endowed with its own nature. However, we must be careful how we understand 'nature' in this context. When Aristotle refers to 'the

[25] This way of reading *GA* II 1 raises a question about how to reconcile Aristotle's two perspectives on generation, since *DA* II 4 clearly refers it to the individual's own good. One answer is to point to Aristotle's distinction between things that exist both for their own sake and for the sake of something else, and those that are present purely for their own sake (e.g. *NE* I 7, 1097a23–34). In *Laws* X Plato uses this distinction and says that each part of the universe is present *both* for its own sake *and* for the sake of the universe as a whole (903b5–d1). Likewise, Aristotle may be saying that biological generation exists *both* for the sake of the individual (*DA* II 4) *and* for the sake of the universe as a whole (*GC* II 10). An alternative answer is to take the two passages to be focused on biological generation, not from two different perspectives, but at two separate levels: *DA* II 4 is concerned with why any particular organism engages in reproduction, while *GA* II 1 is concerned with why biological generation occurs eternally. And there is no reason to think he gives the same answer to both questions.

[26] The argument that follows comes from Henry (2013).

nature of the whole' in the *Metaphysics* passage, there can be no doubt that he is referring to the order and structure displayed by the universe (cf. *Metaph.* V 4, 1015a10–11; *GC* II 9, 335b4–7) and not to some cosmic agent that imposes that order on it like Plato's Demiurge. What I find objectionable about Sedley's interpretation is the idea that Aristotle thinks of the parts of the universe as somehow being adjusted to one another in such a way that their mutual interactions contribute to the cosmic good and, ultimately, the good of man. There is so little evidence for this interactive dimension in Aristotle's natural teleology that I find it hard to believe this was a core feature of his theory. If it were, then we should have expected Aristotle's biology to be driven by a deep interest in ecology and ecological relations. Yet, we find no traces of the concept of an ecosystem, no sign of the idea of 'the web of life', and very little attention to the ubiquitous co-adaptations that exist between living things. (This lies in stark contrast to the ecologically rich perspective of Darwin's *Origin of Species*.)

Sedley's main response to this objection is to say that, while Aristotle's biology is 'squarely focused on individual bodily functioning' the global teleology is supplied by metaphysics. I have two replies to this. First, it is not that the biological works lack examples of such 'coordination' between species (see, e.g., *PA* IV 13, 696b24–35, *GA* III 10, 760a31–b1, *GA* IV 10, 777b16–778a9). The problem is that there are so few of them. If Aristotle really did view nature through the lens of Sedley's interactive teleology, then we would expect his biological works to be full of such examples. Second, it is not clear to me why a study of the coordination and interactions between the parts of nature should belong to metaphysics and not to the science of nature itself. For that is essentially the science of ecology. The absence of anything like an ecological perspective from Aristotle's science of nature and the fact that his biology is so squarely focused on individual bodily functioning makes much better sense if we assume that he rejected Plato's cosmic teleology.

Necessity and eternity

To close this chapter, I want to turn to *GC* II 11 and consider the relation between necessity and the continuity of biological generation. Aristotle's discussion in this chapter is extremely challenging, and its arguments are not easy to decipher. My aim here is not to provide a decisive reading but simply to motivate a particular interpretation.[27]

[27] For a similar reading see Lennox (1985, 73–76).

At the outset of the chapter Aristotle says that every continuous process of change involves consecutiveness where one thing comes to be after another with no interval between them (337a35–b2). The aim of the chapter is to investigate, in the case of coming to be, whether any members of the series are absolutely necessary or whether they are all merely contingent in the sense that they could fail to come into being:

> The question is whether all <generated> things are of this sort or whether for some of them it is necessary in the strict sense (*anankaion haplôs*) that they come into being. In other words, is it the case that, just as in the sphere of being some things cannot possibly fail to exist, so too concerning generation [sc. some things cannot possibly fail to come into being]? For example, is it necessary that the solstices will occur in the sense that it is not possible for them to be incapable of occurring? (337b10–13)

Joachim calls *GC* II II an 'appendix' because he thinks the main task of *GC* is completed in *GC* II 10 and that the discussion of necessity in the final chapter adds nothing to this. I think this is a mistake. The discussion of necessity is clearly connected to *GC* II 10 and the theme of continuous generation. As I read the chapter, Aristotle is ultimately worried about the possibility that biological generation may not continue eternally but could cease at some point. Aristotle will go on to argue that in every process of generation it is never necessary in the strict sense that the end product will come to be; its generation is always contingent. This is especially true in the case of living things, where the generation of each individual is dependent on a number of complex factors that are themselves contingent (e.g. one or both of its parents might have been infertile, or they might have failed to conceive, or its development might have been derailed due to some accident in the process, and so forth). But if it is possible that each individual might fail to come into being, then what is the source of that necessity that is supposed to make biological generation itself eternal? What ensures that there will always be a *genos* of humans, animals, and plants?[28]

The argument of *GC* II II proceeds in three parts. (1) Aristotle first presents a series of arguments to show that it is not necessary in the strict sense for any generated substance to come into being; each one is such that it could have failed to come to be (337b14–33). Its coming to be is only 'conditionally' necessary, as Aristotle puts it.[29] From this he concludes that

[28] Although *GC* II II starts off as a general worry about the necessity of all generated substances, by the end of the chapter it becomes apparent that Aristotle's main concern is the eternity of biological generation (338b6–12, translated below). See also Lennox (1985, 76–7) and note 32 below.

[29] To say that something is 'conditionally' necessary is to say that it must come to be *if* such-and-such is to be or come to be (e.g. the foundation must come to be *if* a house is to be; for it is presupposed by

what is necessary in the strict sense must occur eternally and vice versa (337b33–338a3; cf. *NE* VI 3, 1139b20–24).[30] (2) Next Aristotle argues that it is in circular motion, and in cyclical generation in particular, that such absolute (*haplôs*) necessity is to be found (338a3–18).[31] (3) Finally, he raises a worry about the generation of living things:

> But then why do some things obviously come to be in this way (as, for instance, rain and air come to be cyclically so that it must rain if there is a cloud and, conversely there must be a cloud if it is to rain), while humans and animals do not return upon themselves so that the same individual comes to be for a second time (for although your coming to be presupposes your father's, his coming to be does not presuppose yours)? Why, on the contrary, does this sort of generation seem to proceed in a straight line? (338b6–11 translated after Joachim)

Back in *GC* II 10 Aristotle had argued that only continuous motion is properly eternal and that circular motion is the only kind of motion that is genuinely continuous. However, a rectilinear process, like natural generation, can 'imitate' circular motion by reverting back on itself thereby forming a cycle (think of the rain cycle where the elements are reciprocally transformed into one another). Here Aristotle is worried that the generation of living things is not cyclical in the right way. This is a problem because he has just argued that what is necessary *haplôs* is eternal and that it is only in cyclical generation that this sort of necessity is to be found. What Aristotle needs to show, then, is that the generation of living things is necessary in the strict sense while respecting the fact that each individual organism is only conditionally necessary and could fail to come into being (cf. 337b26–7). In a word, he needs to show how *eternal* lineages can be composed of a series of *contingent* beings. Let me flesh this out a bit, since this is not the traditional way of reading the chapter.[32]

the existence of a house). See Cooper (1982). *GC* 337b14–33 provides a set of complicated arguments to show that, in any causal sequence where some prior member *A* is the cause of some subsequent member *B*, both *A* and *B* will always only be conditionally necessary. Joachim (1926, 272–4) offers a good exposition of those arguments.

30 Cf. Charles (1988, 14–17).

31 Joachim (1926, 273) summarises the doctrine thus: 'No member of a rectilinear succession of *gignomena* . . . can exhibit "absolutely necessity of occurrence". If a thing is to come-to-be with "absolutely necessity", it must come-to-be always and invariably: and that is possible only if it is a member of an eternally-repeated cycle of *gignomena* (37b33–38a5). Hence "absolutely necessity of occurrence" and "reciprocal necessary *nexus*" (which depends on it) are to be found only in cyclical *kinêsis* and cyclical *genesis* (38a5–17).'

32 My interpretation, which takes Aristotle to be worried primarily about the fate of biological generation, is suggested by the puzzle at 338b6–12 (towards which the entire chapter appears to have been building) where the relevant contrast is between meteorological phenomena and heavenly motions, on the one hand, and biological generation, on the other. There are two reasons for thinking Aristotle is not concerned with the continuity (and therefore eternity) of elemental motion in

It is often supposed that Aristotle thinks biological generation is *only* necessary in the conditional sense. Leunissen, for example, argues that absolute necessity holds for eternal things such as the movement of the heavens while 'conditional necessity holds for the generation of animals, which is a sublunary natural process that is rectilinear and that concerns beings whose substance is perishable'.[33] Obviously Leunissen takes Aristotle's statement about biological generation at 338b6–11 to express his settled position in *GC* II 11. But, first, we should note that Aristotle only says the process seems to be (*eoiken*) rectilinear (338b11). I take this to be a tentative statement that is part of the set-up for the problem in question. Second, it is important to distinguish between the claim that the generation of some particular organism (e.g. Socrates) is only conditionally necessary from the claim that the phenomenon of biological generation itself is only conditionally necessary. Now Leunissen is right that Aristotle thinks absolute necessity holds only for eternal things, while the generation of *each particular* organism is a rectilinear process that involves only conditional necessity. Aristotle seems to say as much at 338b9–11 (cf. 337a25–29). But what Aristotle wants to say (and, in fact, goes on to say, if I am reading the text properly) is that *biological generation itself* is a cyclical process and therefore necessary in the strict sense. As we have seen, the *GA* II 1 passage contains an argument for why the generation of living things will continue forever so that there will always (*aei*) be a *genos* of humans, animals, and plants (731b35–36). And, according to *GC* II 11, what occurs always is necessary in the unqualified sense. Herein lies the paradox. Aristotle thinks that generation of every particular organism is not absolutely but only conditionally necessary so that each one could have failed to come into being. But if so, then it should be the case that biological generation itself could cease at some point.

Now we have seen that Aristotle thinks only continuous motion is properly eternal and that only circular motion is genuinely continuous. But he also holds that rectilinear motion can 'imitate' circular motion by being cyclical (i.e. reverting back on itself). And it is in cyclical generation, he says, that absolute necessity is to be found. He says that, at first glance,

GC II 11. First, Aristotle has already laid it down in *GC* II 10 that elemental transformation is a continuous process of generation in which the elements 'return upon' themselves and thereby 'imitate' circular motion (337a1–7). Second, the *GC* II 10 passage appears to have the rain cycle in mind, which is one of those processes of generation that 338b6–12 takes to be 'obviously' continuous/cyclical. For this reason I take *GC* II 11 to be raising a new worry about the continuity (and thus eternity) of biological generation. While it is true that Aristotle mentions the elements in the final paragraph of the chapter, he does so only to illustrate the kind of solution he wants to apply to the case of living things. The point is to extend what he said about the elements in *GC* II 10 to this central case (living things must, like the elements, return upon themselves, not numerically, but formally).

33 Leunissen (2010, 105).

it seems as though (*eoiken*) biological generation is not cyclical in the right way; for 'animals do not return upon themselves so that *the same individual* comes to be a second time' (338b8–9). Once a particular animal passes out of existence, it is gone forever. By contrast, it is numerically the same sun that returns again and again to the same position every time it revolves around the earth. This is where I think many commentators leave the argument. But Aristotle continues by distinguishing between two different ways in which a rectilinear causal process might be cyclical and so imitate circular motion:

> In discussing this, we must begin by inquiring into whether all things return upon themselves in the same way or whether, though in some sequences what recurs is *numerically* the same, in others it is only *the same in form*. Of those moving substances that are imperishable, it is clear that <what returns upon itself> will be numerically the same (for the motion is consequent upon that which is being moved). But those things that are perishable must return upon themselves formally but not numerically. This is why, when water comes to be from air and air from water, the air <that returns> is the same in form, not the same in number. And even if these are numerically the same, this, at any rate, is not the case with those things whose substance comes into being, being the sort of thing that is capable of being otherwise. (338b6–19 translated after Joachim)

Aristotle obviously thinks biological generation is an instance of this kind of formal replication where what returns is not numerically the same individual but something that is the same in form. In this way biological generation itself takes on the character of a cyclical process. And since all cyclical processes are necessary in the strict sense, and since what is absolutely necessary is eternal, it follows that the change must be eternal and continue without fail. Admittedly Aristotle's conclusion is ultimately disappointing because he does not make clear why cyclical processes are necessary in the unqualified sense. As far as I can see he still owes an account that connects the idea of being cyclical to the sort of necessity that guarantees the process will continue eternally without fail. More work needs to be done on this point. But I leave that for sharper minds than mine.

CHAPTER 7

The two kinds of end in Aristotle
The view from the de Anima

Thomas K. Johansen

Aristotle says in many places that the end or goal (*telos*) is a cause. In some of these places, he also warns us that the notion of a *telos* is ambiguous. A *telos* may mean the end for the sake of which a change or activity takes place, but it may also mean the beneficiary of the change or activity. Linguistically, Aristotle marks the distinction by the genitive *to hou heneka* vs. the dative *to tini*. The genitive suggests an object of pursuit, often as an object of a verb for desire (such as *oregetai* or *ephietai*), while the dative is naturally taken as a dative of interest. I shall refer to the one as 'the end-genitive' and the other as 'the end-dative'. Ancient commentators illustrated the distinction by the example of medicine, for the sake of health and with the patient as beneficiary.[1]

The two kinds of *telos* in the *Physics*

Teleology is of course a central tenet of Aristotle's philosophy. So one would think that a distinction between two kinds of *telos* was of basic importance to our understanding of Aristotle's philosophy. Yet until recently the distinction has been met with general indifference.[2] So Hicks comments on its first occurrence in *de Anima* II.4 (415b2):

> This is of the nature of a footnote. It is repeated below 415 b 20. Probably either here or there it is out of place. We find it again, unnecessarily interrupting the argument, in *Metaph.* 1072 b 2–6; *de Gen. An.* II. 6, 742 a 22 sqq., *Phys.* II. 2, 194 a 35 sq., *Eud. Eth.* VII. 15, 1249 b 15.[3]

Ancestors of this chapter were delivered to the 'Explaining Nature' conference at the University of Patras and the Oxford Workshop in Ancient Philosophy. I am grateful to members of those audiences and to Istvan Bodnar and David Ebrey for much constructive criticism.

1 Simplicius, *In Aristotelis physicorum libros commentaria* 303.28–32.

2 Kullmann (1985), Sedley (1991), Menn (2002) and Johnson (2005) deserve credit for refocusing scholarly attention on the distinction.

3 Hicks (1907, 340).

The dismissal of the distinction reflects the view that Aristotle is only serious about one of the senses distinguished, the end-genitive. For it is only in this sense of 'end', it is thought, that ends are operative in nature. So the expectation is that when Aristotle introduces the distinction in these passages he is only doing so to set aside the dative use, and to reassure us that we are talking about the *telos* in its proper genitive use.

Now two of the five occurrences of the distinction occur within one page of *DA* II.4. It is these passages I want to focus on in the bulk of this chapter. I want to argue that while the end-genitive serves to define the characteristic capacities of the soul, there is a significant use of the end-dative to explain other activities of the soul. But before attempting this argument, I shall consider the passage in *Physics* II.2. For the context is programmatic for the role of teleology in Aristotle's natural philosophy, to which of course the psychology belongs. So we can hope to receive a steer from the *Physics* on how to understand the distinction in the *de Anima.*

> Again, 'that for the sake of which,' or the end, belongs to the same department of knowledge as the means. But the nature is the end or 'that for the sake of which.' For if a thing undergoes a continuous change and there is a stage which is last, this stage is the end or 'that for the sake of which.' (That is why the poet was carried away into making an absurd statement when he said 'he has the end for the sake of which he was born.' For not every stage that is last claims to be an end, but only that which is best.) For the arts make their material (some simply 'make' it, others make it serviceable), and we use everything as if it was there for our sake. (We also are in a sense an end. 'That for the sake of which' has two senses: the distinction is made in our work *On Philosophy.*) The arts which govern the matter and have knowledge are two, namely the art which uses the product and the art which directs the production of it. That is why the using art also is in a sense directive; but it differs in that it knows the form, whereas the art which is directive as being concerned with production knows the matter. For the helmsman knows and prescribes what sort of form a helm should have, the other from what wood it should be made and by means of what operations. In the products of art, however, we make the material with a view to the function, whereas in the products of nature the matter is there all along. (*Phys.* 194a27–b8; revised Oxford translation)

One reading of this passage, which we find in Konrad Gaiser,[4] says that Aristotle wants to distinguish the sort of *telos* we find in nature, the end-genitive, from the sort we find in the human sphere but not in nature, the end-dative. Ends in nature are intrinsic to the processes that produce them,

[4] Gaiser (1969, 106).

while human beings use nature for their own sake in ways that are extrinsic to natural things themselves. Acorns, let's say, grow into oak trees; humans cut them down and make furniture of the wood. The uses we put natural things to have nothing to do with the ends that govern the natural things as such. When Aristotle says in the passage that we are 'in a way' (*pôs*) an end, he means just that we use things to our own benefit; however, our being benefitted is not an end of those things as such. So 'in a way' implies 'in a qualified but not in a strict sense'. The end-dative is, then, only to be found in the context of human agents, and the relationship of the end-genitive to the end-dative is contingent. The analogy between nature and the arts is limited to the end-genitive: there are productive arts and using arts; the productive arts correspond to what nature does in bringing about a certain end, whereas the using arts correspond to what we do with the products of arts. The end-dative is restricted to cases of human agency and there is no equivalent at work in nature.

Now I think this reading both overstates and misstates what the passage is saying. One point is that the distinction between the productive and the using arts is supposed to bring out two different ways the arts relate to matter: the productive art is directly concerned with matter in reasoning about the kind of matter that will best realise a certain form, while the using art is indirectly concerned with it in prescribing the form that the matter is then to realise. However, that this distinction between the two kinds of ends exists in the arts is not an indication that it does not also exist in nature or that it is only because we human beings use natural beings that they in any way have an end-dative. This *may* be so, but the passage doesn't tell in favour or against it. That was the overstatement.

The misstatement is the claim that the relationship between the end-genitive and the end-dative is always *contingent*. Contingency may obtain in the case of some artefacts but not all. The doctor aims at the end-genitive of health, but the relationship between health and the patient, the end-dative, is clearly not contingent. This point about the way the productive art depends on the using art goes back of course to Plato's *Republic* x,[5] where the example was the flute-maker and the flautist. The flautist tells the flute-maker how to make a flute, so the end of the producer's art is set by the user's art. If then we associate the using art with the end-dative and the productive art with the end-genitive, we can hardly maintain as a general claim that the relationship between the two is contingent. Of course, this does not exclude that it may be contingent either in those cases

[5] Cf. also *Euthydemus* 291c–d.

where natural entities are used by humans for their own ends, or in cases, if any there be, where the end-genitive and the end-dative are both features of natural entities. But the point shown by the arts is that this contingency cannot be a feature of the relationship between the end-genitive and the end-dative as such.

One aspect of the distinction that needs clarification here is the relationship between the end-genitive and the good. Aristotle completes the first paragraph of the *Physics* passage by underlining that not every terminus of a process counts as an end, but only the best. Certainly there is no need always to qualify the best by a dative 'for which/whom'. To make this kind of point Gaiser refers to various passages in Aristotle's ethical works where the good simply speaking (*haplôs*) is used to contrast with the good for someone or something (*tini*).[6] But there are two possible kinds of claim involved here: one is that we need not add *tini* when we talk about something as good *haplôs*, which is true; another claim is that it is not implied that what is good *haplôs* is good for some kind of thing, which is often, at least, false. So Aristotle distinguishes between what is good *haplôs* for man, say a life according to excellence of reason, and what is good for some individual or other (eating five pounds of flesh is good for Milo the wrestler, but not for some other man).[7] *Tini* here ranges over individuals within the species, whereas *haplôs* may mean what is good for the species man as such. But if so, it should be clear that Aristotle's distinction between what is good *haplôs* and what is good *tini* says nothing about whether we should understand the good as a good for a certain species of being. It would after all be rather against the argument of *Nicomachean Ethics* I.6 to say that there was a good *haplôs* that wasn't the good for or with respect to some kind or category of entity.

Now Aristotle said in *Phys.* II.2 that the end-genitive is a good. Does he think that the end-dative introduces some further good, or may it, at least in some cases, so to speak help spell out the way in which the end-genitive is good? There are clearly cases where we think that the end-dative concerns a kind of good distinct from the end-genitive. So the acorn's becoming a tree is distinct from the good for us of having wood to make beds out of. In this case, the specification of the good of the acorn won't make any reference to its utility to us. But there are also cases where the end-dative seems to articulate the goodness or at least an aspect of the goodness of the end-genitive. So in the case of health, why is health good? Clearly in

[6] *EE* III.1 1228b18–30, VII.2 1235b30–1236a15; *Nic. Eth.* V.1 1129b3, VII.12 1152b26–33.
[7] Cf. *Nic. Eth.* II.6.

part because it is beneficial to the patient. A doctor tries to make his or her patients better. We might in some cases, say where the treatment would be too painful or the patient is resolved to die, imagine a doctor who abandons the attempt to restore health, for the good of the patient. It is reasonable to say that promoting the good of the patient falls within the remit of the doctor as such; it is not just, for example, a consideration brought on by any general ethical duty to help others.

Sometimes this connection between the end-genitive and the end-dative may not be obvious because it emerges only at the level of a higher end-genitive. So brick-making is subject to the art of house-making, which is for the good of man, but one might well think that brick-making itself does not have a beneficiary.[8] Similarly, it could be that there are processes in nature that have an end-genitive that is subject to another end-genitive whose goodness is to be referred to an end-dative only at this higher level. Think, for example, of the roots that plants shoot, which allow the plant to absorb the nutriments in the ground. The immediate end-genitive of roots' growth might be said to be their downward position, but this process is clearly subservient to the aim of taking nutriment (cf. *Phys.* II.8 199a27–29), a good for the plant.[9] The plant seems here to be an end-dative. Of course we may say that the end-dative is obvious at this level of nutrition given its end-genitive *of preserving the plant*; yet the end-dative still adds the information that this end is *good for the plant*.

Now, to generalise, it is particularly when we consider the *goodness* of the end-genitive that it is tempting to introduce the end-dative. There is a Platonic background to this, as we can see in the *Timaeus*. When Timaeus seeks to explain what the primary cause or *aitia* of the eyes is, he at first refers to it like this:

> We must go on to describe the function that makes them [sc. the eyes] most beneficial, because of which god has given them to us. For I reckon that sight has become the cause of the greatest benefit to us in that not a word of all that is being said now about the universe would ever have been said if we had not seen stars and sun and heaven. (*Tim.* 46e7–47a4)

Clearly the reference to us as the beneficiary is an important aspect of the cause. But Timaeus in the following lines also uses the language of *hou*

8 Alternatively, one might think, as David Ebrey has suggested to me, that there are nested benefits along with nested ends. So in the first instance what is benefitted is the house: it has better walls, while the house in turn benefits the person. One issue with this suggestion is clearly whether we can make sense of something's being good for a non-sentient being, on which see below.

9 Cf. *DA* II.4 416a2–18; III.12 434a22–26.

heneka in a way that seems at least a partial anticipation of Aristotle's use: 'god devised sight and gave it to us so that we might see the revolutions of intelligence in the heavens and use their untroubled course to guide the troubled revolutions in our own understanding' (47b6–8). Timaeus' claims are fairly rendered by saying there is an end-genitive of vision, seeing the revolutions in the heavens and correcting our reasoning, but also an end-dative, us for whose benefit god devised sight.

Now it is no doubt relevant that Plato premises the cause (*aitia*) on a *craftsman's* providence: the function is good because god gave the eyes to us for that reason, to be for our benefit. In the context of this kind of explanation it is no surprise that the goodness of the end-dative arises immediately in the context of explaining the end-genitive. However, there is a way of putting this point even without reference to a demiurge: there is a function that benefits the organism that has it. It is when we start thinking of ends-genitive as good in this way that it becomes hard to avoid the beneficiary, the end-dative.

However, it may well be another point introduced by this Platonic background that makes scholars wary of assigning much work in Aristotle's natural philosophy to the end-dative. Isn't the end-dative a way of smuggling in a further end in nature, which allows for a cosmic anthropocentric teleology?[10] So in the *Timaeus* it is not just the eyes that were made for our benefit, but also the sun.[11] However, this fear is unfounded. To say that we use natural entities is clearly not to say that natural entities themselves are for the sake of us – that nature is such as to be for our benefit. It's just to say that the arts pick up things in nature and use them for our good, and in these cases we are the end-dative. But it does not follow from this fact about the arts, that it uses natural beings for the benefit just of man, that if there were also an end-dative in *natural* causation, that is, an end-dative outside of artistic uses, then the beneficiary would always be the same, man or some other thing. What does seem to follow, however, is that if there is also an end-dative in nature it is reasonable to think of it as standing as a user in relation to the end-genitive in such a way that it benefits from its use.[12]

As is well known, there is a passage later in *Phys.* II.7 that is naturally read as restricting the use of final causes to particular substances:[13]

[10] Cf. Sedley (1991, 180). [11] See *Timaeus* 39b–c.

[12] This is not to exclude that there will also be ends-dative that use features that do not themselves have ends-genitive. The omentum discussed below is one such feature.

[13] For this reading see Judson (2005).

> We must explain the 'why' in all the senses of the term, namely, (1) that from this that will necessarily result ('from this' either without qualification or for the most part); (2) that 'this must be so if that is to be so' (as the conclusion presupposes the premises); (3) that this was the essence of the thing; and (4) because it is better thus (not without qualification, but with reference to the substance in each case). (*Phys.* 198b5–9, revised Oxford translation)

(4) seems to be the specification we want for the end-dative in cases such as *eudaimonia* for human beings and nutrition for plants. In these cases the end-dative is the same thing as the subject of the end-genitive. But what the end-dative adds is that the end-genitive is good for a certain kind of thing.

This is a feature that we shall find again in the cases of the end-dative that Aristotle uses in his psychology: the end-dative is the same entity in which the end-genitive is realised, not a distinct entity. There is an analogy between the end-dative in natural beings and artefacts: in both cases the end-dative refers to a user that benefits from the end-genitive. But the ends-dative in the two cases also differ in a way that parallels the way art and nature differ with respect to the efficient cause: both the efficient cause and the end-dative are internal in natural beings, and both are external to artefacts. The role of the end-dative in natural causation, so understood, is not one that supports anthropocentric cosmic teleology since this would make the end-dative, man, external to many of the things of which it is the cause.

To sum up, so far: there appears to be a way in which the end-dative helps explain the works both of art and nature. This becomes apparent particularly when we consider the end-genitive as good, where the end-dative helps specify the respect in which it is good. The difference in such cases between the end-dative in the arts and in nature is that it belongs to the natural being itself, not to an external entity such as man.

The distinction in the *de Anima*: the first occurrence

I turn now to the *DA* to show how the end-dative helps explain psychological phenomena. The distinction between the two ends occurs twice within one page of *DA* II.4. The first passage, 415a26–b7, introduces the nutritive capacity. Aristotle says that animals have offspring in order to participate in the eternal in the way possible for them. For they all desire this and do everything they do in accordance with nature for the sake of this, and then he adds 'that for sake of which is said in two ways, as that at which it aims (*to hou*) and as that which benefits (*to hôi*)'.

It is the goal-directedness of reproduction that occasions the distinction between the two ends. Participating in the eternal seems, as the object of desire, to be the end-genitive of reproduction. But it is hard to see how the end-dative applies. Surely it is not for the benefit of the eternal or god. And how could it be for our own sake? For we as individuals are not around to enjoy our own afterlife, and the species, which does survive, does not seem to be of the sort that can enjoy anything.[14] Given this difficulty, one might conclude that the distinction is simply there to prevent any impression that the Platonic picture is in play, that is, to block the introduction of the end-dative. To all intents and purposes we could do without the distinction: Aristotle is interested only in the end-genitive. Trendelenburg was perhaps right to recommend excision.[15]

However, before moving to such a conclusion, it is worth looking more closely at another Platonic precedent, which everybody recognises as the background to this passage, namely Diotima's speech in the *Symposium* (206a–209e). Diotima claims that lovers desire 'the good to be theirs forever' (*to agathon hautois... aei einai*, 206a6–9). The dative here is possessive, but the passage clearly suggests that we are interested in this as a good for us. The *ergon* or function of love, meanwhile, is procreation in the beautiful. It aims at producing 'another thing such as it was' (208b). And this point of the *Symposium* is clearly echoed in the passage before us when Aristotle uses the phrase 'to make another thing such as oneself' to describe the end-genitive of reproduction. However, this description does not show why mortal beings are interested in reproduction. This becomes clearer when we describe the end as participating to the extent possible in the divine and the eternal. For divinity sounds like something worth having for a mortal being. Once we describe the end-genitive of nutrition in this way, in a way that advertises what is valuable about the process, it seems appropriate also to specify mortal beings as the end-dative.[16]

The problem remains how a mortal being could be the beneficiary of perpetual existence. There are different ways the dilemma could be diffused. Aristotle does seem to think in *Nicomachean Ethics* 1.10–11 that an individual can benefit or be harmed by what happens to its descendants, but it is doubtful whether the limited dependence of one's happiness on the success of one's offspring allowed for here would motivate a concern with the eternal. However, one might think there is a way in which the species,

[14] Cf. Johnson (2005, 76).

[15] Importune intrusa; repetuntur infra, ubi suum locum habent, unde huc manasse videntur.

[16] This indeed is the kind of reading adopted by Themistius, Simplicius and Philoponus. See the helpful discussion in Johnson (2005, 67–69), who appears to agree with this interpretation.

despite its eternity, is capable of being benefitted in that its continued existence, as *Phys.* VIII.6–7 tells us, involves a process of continued generation that ultimately relies on an external cause, the first unmoved mover. In contrast, the eternal existence of the first unmoved mover involves no change and no external causes of its being, and for that reason cannot be benefitted by any changes. Since, then, the species of living beings require external help to ensure their continued existence, the obstacle that precludes the first unmoved mover's being a beneficiary does not apply to them.[17] Of course we may still think that a species is not a proper subject for enjoying anything, but (a) we should be careful not to assume that only sentient subjects can 'enjoy' benefits in the sense relevant for Aristotle – clearly plants (which have no sentience) are beneficiaries of reproduction[18] – and (b) one may think of the species distributively and say that the species enjoys immortality in the sense that each and every one of its members enjoys its own existence.[19]

The distinction in the *de Anima*: the second occurrence

I turn now to the main act, the occurrence of the distinction in *DA* II.4 415b8–22:

> The soul is the cause (*aitia*) and principle (*arkhê*) of the living body. But these are said in many ways, and similarly the soul is cause in three of the defined ways: for the soul is the cause both as that from which the change is and as that for the sake of which and as the substance of ensouled bodies. That it is the cause as substance is clear: for the substance is the cause of being for all things, and living is being for living beings, and the soul is the cause and principle of this. Moreover, the actuality is the formula (*logos*) of what is potentially. It is clear that the soul is a cause also as that for the sake of which. For just as the intellect acts for the sake of something, in the same manner does nature, and this is its end. The soul by nature is such an end in animals. For all the natural bodies belonging to the soul are instruments,[20] the bodies of animals, like the bodies also of plants [are instruments], being for the sake of the soul. For 'that for the sake of which' is said in two ways, as that of which and as that for which. And the soul is also [that] from which locomotion proceeds.

Now when Aristotle once more mentions the distinction between the two kinds of *telos* we may think, again, that it is a knee-jerk reaction to the

[17] See *EE* VII.15 1249b9–16.
[18] As Johnson (2005, 66 n. 7) rightly notes vs. W. D. Ross (1924, 376).
[19] For the teleology of animal generation, see further Devin Henry in this volume.
[20] 'All the natural bodies are instruments of the soul' may be the more obvious translation, but it is not the kind of thing Aristotle would want to say or want to say here.

mention of the final cause. But when we look at what comes before the distinction, there seems good reason to think that the beneficiary cause is an integral part of Aristotle's thinking. So he points first to the way *nous* makes things by analogy with nature. Intelligent production suggests the case of artefacts that have not just an end but also a beneficiary, since artefacts generally involve users and beneficiaries. Similarly, we might think that nature has created the bodies of living beings so that they can use them for their own benefit. Aristotle's thinking may, then, be close to that of *Phys.* II.2 where he used the arts to indicate the presence also of an end-dative. Accordingly, Aristotle here underlines the instrumental nature of the body where again we think that instruments like hammers and saws (the sorts of instruments Aristotle directed us to in II.1) have not just an end function, but also a beneficiary. We would think then of the soul not just as the fulfilment of the body in the sense that its capacities and activities realise the body's potential as an instrument, that is, think of the soul as an end-genitive, but also that the soul benefits from the body in that it uses parts of the body in realising its characteristic activities.[21] We need not separate the soul from the body to make this point: we can think, for example, of the whole living being using a part of its body to realise a certain function. By analogy, a corporation may use an employee, who is part of the organisation, to execute a certain job, which is good for the corporation as a whole. Or we might say that one part of the organisation, the administration, for instance, uses another part of it to fulfil its ends. There need be no contradiction for Aristotle in the notion of the living being using a part of itself for its own benefit.

Is it right, however, to think of the soul in this case as the end-dative rather than the living being as a whole? One might think that the soul as such does not have its own interests and benefits just for the reason that the soul is not a substance in its own right. Humans, dogs and trees have interests: it is after all such substances that live and live well, not their souls as such. To avoid the implication that the soul has interests in a manner appropriate only to substances,[22] one may take the sense in which the soul is the end-dative to be as a specification of the respect in which the whole living being is properly said to be benefitted. In *DA* II.4 Aristotle (415b11–14) refers to the soul as being most properly the substance of the

21 See the helpful discussion in Menn (2002, 110–14).

22 Menn (2002) argues that Aristotle presents the soul as being like the art rather than the artisan in order to avoid the implication that the soul is a distinct substance in relation to the body it uses. But it is not clear to me how (if at all) Menn would avoid the impression that the soul understood as the beneficiary is like a substance in having its own interests.

living being, that in virtue of which it lives. If, then, we think of the soul as that in virtue of which the living being is alive, we may also consider that it is the living being in virtue of being ensouled that is most properly in a position to enjoy benefits. The instrumental body helps us live and live well, and living and living well are something we do in virtue of our soul. Similarly, you might think that it is in virtue of being ensouled that a living being is properly able to enjoy good health, because the benefit is tied to the states and processes of living. (Of course the *perception* of well-being will be one of those but there is no need to limit the benefit to this life function.) Given then that the benefit of the body accrues to us in virtue of living, and living is a function of the soul, it makes sense to think of the living being qua ensouled as the beneficiary. Overall, we might prefer to say, in line with 408b13–15,[23] that the living being qua ensouled uses its bodily parts to realise its functions.

Psychological applications of the two ends

So the claim is that the soul, or the living being qua ensouled, is the end-dative of the body considered as its instrument.[24] The soul uses the bodily parts for its own benefit rather in the manner a craftsman uses his instruments for his own ends.[25] The difference is that the living being's instruments are part of it, not distinct as are the craftsman's tools. This, however, is not the end of Aristotle's interest in ends-dative.

What I want to suggest is that the end-dative is a handy device for Aristotle when he seeks to explain teleologically a wide range of things we do in virtue of our soul. I shall argue that while it is the end-genitive that serves to define the various capacities of the soul that define the soul as such,[26] there are also uses of these capacities that do not themselves have ends-genitive though they can be explained in relation to the living being as the end-dative. The use of the end-dative presupposes capacities with

[23] 'For it is probably better not to say that the soul pities or learns or reasons, but that the human being does so with the soul.'

[24] 'The living *being* qua ensouled' might in some contexts be distinguished from 'the living *body* qua ensouled'. So in nutrition (416b10–30) Aristotle says that the nutritive faculty uses food to nourish the living body qua ensouled. Here, while only living, ensouled beings are nourished, it is clearly important to specify what is nourished as a body, since it is as a *quantity* that needs to maintain its form that it is nourished.

[25] As Menn (2002, 113) puts it, 'the body is for the sake of the soul as τὸ ᾧ, the-benefit-to-whom, as an ὄργανον is for the sake of the art or the artisan'.

[26] I argue for this claim in Johansen (2012). For a contrasting view, which denies that capacities define the soul, see Christopher Frey's contribution in this volume.

ends-genitive but allows us to extend teleological explanation in the *DA* to psychological activities not captured by the end-genitive.

Consider the capacity to perceive. The capacity to perceive is the formal cause of the sense-organs. It is the presence of the capacity to perceive that makes the sense-organs *such*. But prior in account (*logos*) to the capacity is the activity of perception, and again prior to this is the object of perception, the sensible form. If you really want to know what a sense-organ is you have to understand what it is to be able to *perceive*, and perception is the reception of sensible form.[27] Our sense-organs are instruments whereby we are able to engage in perception so defined. Formal and final causation coincide in the account for the sense-organs: the activity for the sake of which we have sense-organs is also the activity that ultimately defines them, defines them, that is, via the definition of the capacity to perceive. We meet again the characteristic feature of the natural philosophical accounts: formal and final causes coincide.

Now I want to show how Aristotle's account of perception also involves a final cause, in the sense of an end-dative. The end-genitive of sight, say, is the actual perception of the proper object of sight, colour. This is what sight is *for* and how we define sight as a capacity. However, it is clear that Aristotle also thinks that sight has further functions. So in *DA* III.12–13 he argues that vision, and the other distal senses, are necessary for the survival of animals that move around, to avoid dangers and spot food and drink from afar. However, the distance senses also contribute to our well-being. This claim is further developed in *Sens.* 1 where Aristotle points to the ways in which sight and hearing contribute to our knowledge. Now it is clear from what I have already said that the end-genitive of sight is not realised in locomotion or knowledge as such: the capacity of sight is for the sake of seeing colours and this is the activity that defines it. Seeing may in turn be *used* in our perambulations and cogitations, but we do not make reference to locomotion or knowledge when defining what sight as such is.

While these passages show that perception contributes to moving and thinking it does not yet follow that the way to think teleologically about this contribution is to bring in the end-dative. For we might think that this is a case of an end-genitive being subordinate to a higher end-genitive, rather as the way the art of brick-making served the art of house-making. This perspective too could accommodate the holistic implications of the cooperation of the various capacities, when we say that *certain animals* need distance sense to move. So a variety of arts come together in the building

[27] See *DA* II.4 415a14–23.

of a house, the arts of brick-making, plastering, plumbing and so on, each with its particular defining end, but all contributing to the making of a house. Similarly, the various functions of the soul might come together with the soul as a whole being their overarching end. Such a picture can be maintained simply in terms of thinking of ends-genitive, one might say. There is no need yet to bring in the end-dative.

However, if this holds for the capacities of the soul, we would expect there to be a closer definitional and explanatory connection between the lower and the higher ends than appears to obtain between the capacities of the soul. Bricks are after all made for the sake of houses, so it seems that the account of brick-making is not complete, to say the very least, unless house-making has been brought in. Similarly, the account of bridle-making is not complete without reference to the use of bridles in horseback riding. Perhaps we do not need the connection between the lower and the higher art to be very specific: there may be arts that prepare rather general-purpose products, such as rubber or steel. However, not even such a general-purpose connection seems to obtain between the capacities of perception and locomotion. Perception *may* in some animals contribute to locomotion, but the account of perception as such makes no reference to *any* further use, let alone locomotion or knowledge; and it couldn't make such reference, given that the account has to apply to creatures that possess no such further functions.[28] The definitional independence of perception from locomotion or knowledge suggests, then, that it cannot be understood as having a further end in either of these that functions as its end-genitive.

Moreover, it is clear that *DA* III.12 uses the dative to pick up on what is good for the organism. This comes in extension of the claim that nature does nothing in vain:

> Why should it not have perception? It would have to be better either for the soul or for the body; but in fact it is neither – for the absence of sensation will not enable the one to think or the other to exist better. Therefore no body which is not stationary has soul without sensation. (434b5–9; revised Oxford translation)

The notion of well-being, as expressed by the good (*to eu*) plus the dative, seems exactly to introduce the perspective of the end-dative, the perspective of what is good for a certain kind of substance. This is also why the teleological explanation involving the same capacities can differ for different

28 The point I'm making here is specifically about the instrumental relationship between ends-genitive; this does not tell against there being explanatory connections between ends-genitive and ends-dative of the sort we observed in the case of medicine between health and patient.

substances; what is necessary for one is also for the well-being of another. The teleological perspective goes beyond the ends-genitive that apply by virtue of the capacities themselves.

Compare the discussion in *Sens.* 1 (436b18–437a17). Aristotle here distinguishes between, on the one hand, sight's contribution, by its proper function of perceiving differences in colour, to our knowledge of the world and, on the other hand, the way hearing contributes *per accidens* to knowledge, insofar as it allows us to hear speech. It seems clear that what sight does as such is something all sight does, animal or human, but also that humans use colour differences, given that they have intellect, to extract other information about the world. In contrast, the proper object of hearing, that is, sound, doesn't as such tell us much about the world. *Speech* does, but it is accidental to sound as such that it makes up speech. Aristotle's point here is not that vision as such contributes to knowledge, but that what vision as such is of, colour, contributes to knowledge for human beings, while hearing neither contributes as such to knowledge nor has a proper object, sound, which is such as to contribute to knowledge. In both these cases, if in rather different ways, it seems attractive therefore to say that the final causes involved are not causes that belong to the visual or auditory capacities as such, rather hearing and sight serve a purpose insofar as they are used in a certain way by human beings for their benefit.

These instances point, then, to a significant role for the end-dative when Aristotle explores the functions of psychological capacities beyond those that serve to define them as such.[29] This is no doubt part of the reason why such ends-dative occur *at the end* of the *DA*, and then in the *Sens.* after the capacities have been defined by their proper functions. The ends-genitive we looked at earlier in the *DA* attached to the capacities of the soul themselves, and made no reference as such to the whole soul or living being to which they belonged. And that is why Aristotle in *DA* II.5–11 could define the perceptual capacities, for example, in the same way for all animals. However, the good that perception served had to be specified in different ways for different animals. These goods are not involved in strictly defining the capacities of the soul as such, but rather emerge when we consider, more holistically, how these capacities are employed by various kinds of living beings for their overall good.

The similarity between the end-dative in art and nature can now be restated. In the arts we have the distinction between the producer's and the

[29] The opening of *Sens.* 1 alerts us to the distinction: Ἐπεὶ δὲ περὶ ψυχῆς καθ' αὑτὴν διώρισται πρότερον καὶ περὶ τῶν δυνάμεων ἑκάστης κατὰ μόριον αὐτῆς, ἐχόμενόν ἐστι, *ktl* (436a1–2).

user's arts. Here the end-dative is linked to the user's art. This art tells the user how to use the artefact to her benefit. In the psychological case the end-dative is the organism as a whole. The organism uses the various ends-genitive for its own good. So the end of sight is discriminating colours, and man will, as a natural knower, use this end for his good in acquiring knowledge. Bringing in the end-dative, as the whole organism, allows us to see explanatory connections between perception and knowledge. It is when we bring in man as the end-dative that we can properly understand the teleological contribution that perception makes to knowledge.

So introducing the end-dative allows us to understand the cooperation between capacities from the point of view of the organism they serve. But we can go further. It also allows us to give teleological explanations of psychological phenomena for which ends-genitive are not available.

Kullmann offers some helpful examples of the end-dative performing a similar role in Aristotle's biology.[30] There are, he points out, material structures that arise by necessity but which the organism then co-opts for its own benefit. In this case the structures do not themselves have an end-genitive. One such case is the genesis of the omentum, the double fold of peritoneum that covers most of the intestines. This is at first explained as a necessary consequence of the formation of the intestines;[31] once formed the omentum is given the function of improving digestion. But, as Kullmann puts it, 'this function is only secondary'. There are other structures, such as hair and horns (*Parts of Animals* III.2 663b22–24), which arise from material necessity but are then used by the animal to protect itself. Such cases seem to illustrate Aristotle's observation that 'At times nature seems to make use even of remnants for some benefit. But that is not a sufficient reason to inquire what all remnants are for the sake of' (*PA* IV.2 677a15–17).

Now we have already seen reason not to restrict the end-dative to features that arise merely by necessity without ends-genitive. So the nutritive and perceptual organs each have their proper end-genitive, but they are also for the sake of the living being, qua ensouled, as their end-dative.[32] Here it made sense to say that the end-dative made explicit the way in which these ends-genitive were good: so in nutrition the end-genitive was participating in the eternal and divine. I also suggested we could understand the coordination of capacities such as perception, locomotion and knowledge within the individual organism from the point of view of this end-dative: aspects of perception, say, the perception of voice, could be

[30] Kullmann (1985, 173–74).

[31] Ἡ μὲν οὖν γένεσις ἐξ ἀνάγκης συμβαίνει τοιαύτη τοῦ μορίου τούτου (*PA* IV.3 677b22–23).

[32] Kullmann's own taxonomy of final causes would suggest that he agrees with this.

understood teleologically by the reference to the good of the animal, even if such perception went beyond the end-genitive that strictly defined the capacity of perception.

Now in the case of useful remnants such as the omentum the end-dative relates to features that are not essential and do not by nature have an end-genitive.[33] It is an important point about Aristotle's psychology, just as it is about his biology, that there are features without an end-genitive that can be teleologically explained by reference to the end-dative. Mor Segev has recently suggested that dreams can be understood in this manner. He argues that dreams arise from necessary processes that do not themselves serve a purpose, but may be put to use by nature for the good of the organism in question.[34] Dreams are in other words like Kullmann's useful residues.

Now dreams are a specific instance of *phantasia*. Without going into detail, I want here to make a similar point to Segev's about *phantasia* as a whole. Aristotle understands *phantasia* as a movement resulting from the activity of perception (*DA* III.3 429a1–3).[35] It is involved in a range of perception-like activities, such as dreaming, remembering and, generally, entertaining images and representations. Much of this work is not to be explained by ends-genitive. So Aristotle indicates that dreaming *as such* does not serve a purpose.[36] However, *phantasia* clearly also contributes to the way we develop knowledge. Knowledge presupposes memory, a kind of *phantasia*, and as Aristotle says in the *DA* III.7 (431a15–17) it is impossible to think theoretically without entertaining a *phantasma*. Perception strictly understood does not suffice. Actual perception is of the present.[37] It depends on a perceptual object that is immediately present to cause perception. Without *phantasia*, it is not clear how perception could retain and

33 A similar line is taken by Leunissen (2010). She distinguishes between primary and secondary teleology, and associates secondary teleology with further uses of material causes. However, her distinction, on her own admission (2010, 5, n. 11), does not represent Aristotle's distinction between the end-genitive and the end-dative, and I shall not try to assess it here, other than to say that I would take it that Aristotle's end-dative does not coincide with her secondary teleology insofar as the latter only applies to what has come about because of material rather than hypothetical necessity.

34 Segev (2012). My own suggestion was made, independently, in Johansen (2012, 219–20). I differ from Segev in that (a) he does not explicate his claim about the usefulness of dreams in terms of Aristotle's distinction between ends-genitive and ends-dative, and (b) he presents dreams as a product of *phantasia* where *phantasia* itself is taken as a distinct faculty with a proper function, whereas I prefer to see *phantasia* not as a faculty in its own right with a proper end-genitive, but rather as a movement arising from perception and so as belonging to the perceptual faculty.

35 The reader will find a fuller account of *phantasia* from this perspective in Johansen (2012).

36 Though, if Segev (2012) is right, dreams may be useful to the organism that has them.

37 See *On Memory* 449b27.

process perceptual information in a way that makes it serviceable to our intellect. But insofar as the perceptual faculty also retains sensory images, as a result of actual perception, we can see how perception can retain perceptual information and make it available to higher-order cognitive processes: memory, experience, skill, practical and theoretical knowledge. Again this point may be best expressed in terms of the end-dative since *phantasia* does not occur *in order* to satisfy these cognitive needs and is not generally defined with reference to its contribution to knowledge. Nor again is perception as such defined with reference to its contribution to *phantasia* or knowledge, since there are animals that have neither *phantasia* nor knowledge, though they must have perception qua animals. It seems better therefore to understand *phantasia* as a process that necessarily arises from perception but is used by different animals in different ways according to their varying needs and interests. The distinction between the two kinds of final cause offers us a way of drawing *phantasia* in under *a* final causal explanation without ascribing an end-genitive to *phantasia* as such.

Consider in the same way, again, the contribution that perception makes to locomotion. We saw earlier how animals that move use perception in a certain way to guide their movements. Now perception strictly defined will only give you access to your immediate current environment. Perception is of the present. But often our movements are directed towards what is not yet present, an object we anticipate or fear. Aristotle refers in the *de Motu Animalium* to *phantasia* as performing an analogous function in movement to perception and thinking (701a34–36). Now if *phantasia* allows us to retain perceptual information, it is possible to see how an animal that moves might employ such information to direct its future actions. A wildebeest might recall drinking water from the Masa and direct its future migrations towards that river. While *phantasia* itself does not have an end-genitive in locomotion, we can say that the wildebeest uses the *phantasmata* in this way because it is an animal with locomotion. Earlier we saw the end-dative help explain the integration of faculties within the organism, where those faculties themselves had an end-genitive. In the case of *phantasia*, we see how functions beyond those that serve to define the faculties can also be explained by their usefulness to the organism as a whole. It may be right to describe the use of *phantasia* by the organism as 'secondary', to use Kullmann's phrase, from the point of view of perception itself, yet it is clearly not secondary from the point of view of the organism. A human being would be not be able realise its nature as rational without the availability of *phantasia*, nor would the migratory wildebeest fulfil its nature.

To conclude: scholars are often sceptical about ends-dative because of their association with cosmic teleology. However, there is a danger here that interpreters in their wish to avoid the coordinating role of ends-dative in the cosmos, also do away with their coordinating role within individual organisms. I hope in this chapter to have shown that there is some reason to think that ends-dative are important to Aristotle's psychology, and that his psychology is of a piece with his natural philosophy and biology on this point.

CHAPTER 8

Two conceptions of soul in Aristotle

Christopher Frey

Aristotle appears to employ two methods when he investigates the soul. What is striking about these two methods is that they yield importantly different, and prima facie incompatible, views about what souls are.

Method 1: To understand life is to understand the various vital capacities (δυνάμεις τῆς ψυχῆς) that are exercised within and by living organisms. In general, to understand a capacity one must first look to the activity that is the capacity's actualization and to understand an activity one must first look to the proper class of objects upon which the activity is directed.

Soul as Capacities (via Method 1): The soul is a multiplicity of independently specifiable and (in some cases) separable capacities that are, in advanced organisms, related to one another in a manner that effects a unity of soul over and above the multiplicity.

Method 2: To understand life is to understand its principle (ἀρχή). Living organisms are natural unities; the principle of their vital movements and activities is a single, unitary nature (φύσις). In general, to understand a nature is to understand the single formal end for the sake of which the activities that have this nature as their principle occur.

Soul as Nature (via Method 2): The soul, a living organism's form, is a nature. This single, unitary nature is the principle and end of all of a living organism's vital activities.

How these conceptions of soul follow from the two methods, the precise ways in which these conceptions of soul differ from one another, and the consequences these differences have for achieving that most fine and prized knowledge that an inquiry into soul promises will become clear as we proceed. But it is important to note right from the beginning that how one

I presented versions of this chapter at Georgetown University, Northwestern University, the Catholic University of America, the University of South Carolina, and Notre Dame University. In addition to the audiences at these talks, I would like to thank David Ebrey, Jennifer Frey, and Sean Kelsey for their helpful comments and criticisms.

reconciles these two conceptions will determine to a large extent what it is for a living organism to be a substance.

The first method emphasizes the complexity of living organisms but has difficulty accounting for their substantial unity. If one begins with each of the capacities individually, it is difficult to see how, once the inquiry is complete, one is left with anything more than a mere collection. Whatever it is that effects a unity among these capacities will be posterior to the capacities themselves and this posteriority appears to preclude living organisms from being proper substances.

The second method emphasizes the unity of living organisms but has difficulty accounting for their complexity. According to the second conception of soul, the various activities that occur in the coming to be and continuation of a life arise from a single, unitary, internal principle of movement and rest and occur for the sake of a single formal end. But how, on this conception, are we to countenance capacities that can be specified in a form-independent manner, a manner that would make it legitimate to describe two kinds of organism, say dogs and humans, as possessing the same part or capacity of soul (e.g. perception)? And more generally, how are we to legitimize talk about vital capacities, parts of soul, and organic complexity at all if souls are unitary?

We can view these two methods as presenting us with an *aporia*. It is a familiar *aporia* in which the demands of unity and multiplicity are at odds. My goal is to defuse the *aporia* in the typical Aristotelian fashion by reconciling the competing accounts of soul these methods yield.

Most contemporary interpreters of Aristotle emphasize the first method. Indeed, Aristotle seems to follow it himself in *de Anima*. He first discusses nutrition, then perception, then thought, and these discussions, as we will see, appear to be largely independent. But the virtues of the second method, in my opinion, have not received the attention they deserve. I aim to show that taking the second conception of soul to be primary puts one in a position to reconcile these opposing views. To defend the priority of the second conception in a manner that serves as a reconciliation rather than the acceptance of one horn of a dilemma over its rival, at least two things must be done. First, I have to show why Aristotle proceeds in the way he does in *de Anima*.

(Q1) Why does Aristotle treat the nutritive, perceptive, and rational capacities in a sequence and provide accounts of these parts of soul that are separable in account?

Second, I have to show how a unitary soul can manifest the complexity Aristotle attributes to it.

(Q2) What is it for a unitary soul to comprise various parts and capacities?

I will first develop these two methods and their corresponding conceptions of soul in turn (second and third sections). These discussions are largely expository; my goal in these sections is simply to present and motivate the opposing conceptions of soul. The heavy lifting occurs in section four, where I defend the second conception's centrality.

I first answer Q1 (in the first part of section four). In *de Anima*, Aristotle is not simply giving accounts of the vital capacities. He is trying to explain what it is *to be* a vital capacity. I will show why it is important to distinguish two questions – (i) For a particular vital capacity, what is it? and (ii) What is it for any capacity, given what it is, to be vital? – and will argue that the second question is central to Aristotle's project. In short, Aristotle aims to reveal what unites the various ways life manifests itself. To do so is to understand what makes them instances of living at all. And without this understanding, neither the soul nor life could be the subject of a single and properly scientific inquiry. I then answer Q2 (in the second part of section four). Aristotle approaches the unity of the ways life manifests itself across species by considering how the hierarchically ordered capacities that underlie these vital activities are united in (relatively advanced) individuals. This individual unity consists in one or more lower souls being present potentially (*dunamei*) in the single, higher soul that is the individual organism's nature. I analyze what it is for a soul to be present *dunamei* in this way. It is this final section that contains the most sustained argument for the conception of soul as unitary nature.

The soul as capacities

The passages that lead interpreters to prioritize capacities in their accounts of the soul begin in *de Anima* II.2. Aristotle says,

> We say, then, making a beginning of our inquiry, that that which has soul is distinguished from that which has not by life. But life is so spoken of in many ways (πλεοναχῶς δὲ τοῦ ζῆν λεγομένου), and we say that a thing lives if but one of the following is present. (413a20–23)[1]

[1] Translations, with occasional minor changes, are taken from Hamlyn (2002) and Barnes (1984).

This is a familiar sort of inquiry. We begin with what is most obvious (φανερώτερος) and proceed to what is ultimately more intelligible and can ground satisfactory explanations (κατὰ τὸν λόγον γνωριμώτερος, 413a11–13). Aristotle takes at least two things to be obvious. First, an inquiry into soul is an inquiry into life, since he simply introduces the soul as the principle of living beings (I.1, 402a6–7; II.2, 413b11–13). Second, we attribute life (and therefore souls) to numerous entities on the basis of what initially appear to be wildly different activities.

Aristotle identifies four different types of activity that we readily recognize as ways life is expressed (413a23–25):

(1) intellect/thought (νοῦς),
(2) perception (αἴσθησις),
(3) movement and rest in respect of place (κίνησις καὶ στάσις ἡ κατὰ τόπον), and
(4) nutrition, decay, and growth (κίνησις ἡ κατὰ τροφὴν καὶ φθίσις τε καὶ αὔξησις).

We take something to be alive, says Aristotle, if it displays one of these types of activity.[2]

Each of the four types of life activity is then said to be the exercise of an underlying capacity or power of soul. Aristotle says that "for the present let it be enough to say only that the soul is the principle of the [activities] mentioned above and is divided into these" –

(1) the nutritive capacity (τὸ θρεπτικόν),
(2) the perceptual capacity (τὸ αἰσθητικόν),
(3) the capacity for thinking (τὸ διανοητικόν), and
(4) movement (κίνησις) (413b11–13).

Though ordered differently, the capacities mentioned here correspond to the previous list of activities that capture the basic ways life is said. The basic life activities are the exercises of these capacities of soul.

At this point, many commentators think Aristotle has settled on thinking of the soul as somehow a collection of capacities for the activities we all recognize as modes of living. Indeed, it is not difficult to find many passages like this:

[2] This is not establishing what life is (or means) but is giving an empirical test. On the basis of what marks could you divide what there is into the living and the non-living? Answer: If something manifests any of these four types of activity, then one can conclude that it is alive. Cf. Matthews (1992).

> Aristotle's statement that the most appropriate account of the soul is the one which picks out these capacities, already suggests the thought that perhaps the soul just *is* these capacities. This thought is confirmed when we notice that Aristotle speaks of the capacities as *parts* of the soul. (Sorabji 1974, 64)

The transition from speaking of capacities as parts of the soul to the claim that the soul simply is these capacities is too quick. Its validity depends on what it is to be a part (μέρος/μόριον) of soul.

One problem with Sorabji's claim is that it does not take into account the fact that Aristotle's use of 'part' is more restrictive than his use of 'capacity' (δύναμις). Aristotle is clear that we must distinguish vital capacities that are parts of soul from vital capacities that are not.

> For those who divide the soul into parts, if they divide and separate them according to capacities, they [sc. the parts of the soul] will become very numerous (πάμπολλα) – the nutritive capacity, the perceptual capacity, the capacity to understand, the capacity to deliberate, and, furthermore, the capacity to desire. (III.10, 433b1–4)

These are not the only vital capacities Aristotle mentions. There is also the capacity to imagine, to remember, to feel pleasure and pain, to dream, and to form opinions (III.9, 432a22–b7). Why not add others still, like the capacity to perform arithmetic, or to skip, or to curl one's tongue, or to pass gas? Without some prioritization, without taking some of these capacities to be more fundamental than others, systematic inquiry into the soul would be impossible and the soul could not serve as a living organism's principle of unity.

But even if we recognize that not every vital capacity is a part of soul, it is still unclear whether Aristotle's willingness to speak of souls as having parts supports the view that the soul is somehow a collection of those fundamental capacities that possess the elevated status of parthood. Some recent attempts to distinguish the soul's parts from its capacities do support this conclusion. And it is important to see how these interpretations of the part/capacity distinction lead to the view that the soul is, in a sense, its parts. I will use the recent and representative account that Corcilius and Gregoric (2010) offer as a concrete example.

According to Corcilius and Gregoric, the soul's parts are "the fundamental aspects of the soul which cannot be explained with reference to one another or to any other capacity of the soul whereas all other capacities can be explained with reference to one or several of them" (2010, 89). That is, the parts of the soul are those vital capacities that are separable in account from all other vital capacities, where separability in account is defined as follows.

> separability in account: x is definitionally independent of y, i.e. x has an account or definition which makes no reference to y. (*ibid.*, 114)

The nutritive, the perceptual, and the thinking capacities satisfy this criterion. The nutritive capacity is a capacity to maintain its possessor as such while food prepares it for its activity (II.4, 416b17–19). The perceptual capacity is a capacity to receive sensible forms without matter (II.12, 424a17–19). The thinking capacity is a capacity to receive intelligible forms or to grasp essential features (III.4, 429a13–17). These definitions do not refer to any other vital capacity. But the accounts of all other vital capacities must refer to at least one of these three.[3]

So, according to this criterion, the number of basic capacities is manageable. But how are we to view their copresence in an individual as more than a mere collection? Aristotle insists that the soul is not a collection; the soul is a unity. How are we to account for this unity?

There are several ways to respond to this question. I will discuss two proposals. The first grounds the soul's unity in the physical overlap of an organism's physiological systems. The second grounds the soul's unity in relations of teleological subordination.

Unity of soul as the physical overlap of physiological systems

In the following passage, Jennifer Whiting suggests that the soul's unity is due, at least in part, to the partial physical overlap of physiological systems that are, with relevant qualifications, functionally modular and thereby separable in place.

> The nutritive and reproductive capacities are embodied in one physiological system . . . while the capacities of perception, imagination, and desire are embodied in a different physiological system. Each of these physiological

[3] For example, the locomotive capacity is not a part of the soul, according to Corcilius and Gregoric, because its account must refer to the perceptual capacity. Aristotle employs other notions of difference and separability in *de Anima*. Corcilius and Gregoric discuss three of these: difference in account, separability in place, and existential separability (*ibid.*, 114). Jennifer Whiting (2002) uses separability in place (which she construes slightly differently than Corcilius and Gregoric do) to determine whether a capacity is a part. On her interpretation, the parts of soul are the nutritive, the locomotive, and the thinking. What is important for our purposes, however, is not which capacities are parts, but an appreciation of the sorts of considerations that are used to identify the parts and the general picture of the soul that results.

> systems is centered in one and the same organ (namely the heart), which helps to explain their unity with one another. But each can (at least in some circumstances) *function* relatively independently of the other. (Whiting 2002, 152)[4]

Cardiocentrism is not the only way in which an organism's physiological systems can overlap physically. For example, Aristotle observes that we can cut some insects into segments in such a way that each segment continues to live. Moreover, each segment keeps all the vital capacities that the insect had prior to being divided (*DA* II.2, 413b16–24; I.4, 409a9–10). Aristotle admits that the physical overlap of an insect's physiological systems is not complete; the segments may not persist for long because the division may exclude some organs that are necessary for self-preservation (I.5, 411b19–27). But the fact that the segments continue to live and are able to exercise all of their vital capacities for any significant duration is evidence that their physiological systems exhibit more physical overlap than our systems do.

But why can't there be organisms with physiological systems that exhibit less or even no physical overlap? Plato, at least in the *Timaeus* (44d; 69c–72d), thinks that *we* are such organisms. The rational part of the soul is located in the head, the spirited in the chest, and the appetitive in the abdomen with no physical overlap.

Perhaps Plato's description of our physiological systems is not really possible. Perhaps cardiocentrism, or some other significant physical overlap, is a minimal condition for any animal's vital capacities to be embodied. But even if the souls of animals and humans require this sort of physical overlap, there is nothing about the notion of physical overlap itself that would explain the soul's unity. Aristotle says explicitly that the only explanation of unity in the vicinity occurs in the opposite direction.

> If, then, its [sc. the soul's] nature admits of its being divided, what can it be that holds the parts together? Surely not the body; on the contrary it seems rather to be the soul that holds the body together. (I.5, 411b6–8)

In fact, if the body (or anything else other than the soul) were "that which makes the soul one, this would have the best right to the name of soul" (411b9–10). There must be something about the soul itself that makes it a unity and there must be something about being ensouled itself that makes an ensouled organism a unity. Even if our physiological systems overlap

[4] To be fair, it is unclear from the passage whether this cardiocentrism explains the unity of the soul's capacities or whether it simply explains a systemic unity that obtains among the embodied physiological systems. It is nevertheless useful for present purposes to view the passage's concern to be the former, i.e. the unity of soul.

physically, to use this fact to explain the soul's unity is to reverse the proper order of explanation.

Unity of soul as teleological subordination

A more promising proposal is that the capacities of soul are a unity because they form a series ordered by the relation 'being for the sake of'. For example, according to Mariska Leunissen, the unity of an organism's vital capacities consists in a bottom-up ontological nesting and top-down teleological subordination. She says "the ontological hierarchy of nested capacities is . . . a 'taxonomical' hierarchy in which the more basic capacities constitute a necessary prerequisite for the existence of the higher and in which the realization of the more complex capacities contributes to the goals pursued by the basic ones" (Leunissen 2010, 59). It is the second condition, the claim that the operations of the ontologically higher capacities occur for the sake of the ontologically lower capacities, that is supposed to effect a unity of soul.[5]

What does it mean for a higher capacity's exercises to be for the sake of those of a lower capacity? Aristotle makes a distinction between two senses in which something can be for the sake of something else. He says, "that for the sake of which is twofold – the purpose for which and the beneficiary for whom (τὸ μὲν οὗ, τὸ δὲ ᾧ)" (II.4, 415b2–3).[6] In its original context, a discussion of the nutritive soul, the application of this distinction is fairly straightforward: the nutritive soul is for the sake of reproduction, its aim and purpose are participation in the divine and eternal, and its beneficiary is the living organism.[7] But we can employ the second sense of being for the sake of, the sense that involves a beneficiary, to relate an organism's vital capacities.

Perceptual capacities in general, and touch in particular, serve as good examples. The possession of perceptual capacities enables one to do more than simply take on perceptible forms. Locomotive animals also exercise these capacities in a way that allows them to gather food more easily and

[5] Relations of teleological subordination can be ordered either top-down, as Leunissen orders them, or bottom-up. Monte Johnson maintains the bottom-up ordering. He says, "[a]lthough the more complex parts and capacities come into being after the simpler, the simpler exist for the sake of the more complex. Thus there is an inversion of the genetic and explanatory orders in the case of organisms" (Johnson 2005, 9). For our purposes, it doesn't matter in which direction the teleological subordination occurs. The very appeal to teleological subordination as a source of unity is what concerns us. To simplify matters, I will focus on the top-down ordering only.

[6] Cf. *Phys.* II.2, 194a35–36; *DA* II.4, 415b20–21; *Metaph.* Λ 7, 1072b1–3; *EE* VII.15, 1249b15; and Johansen (this volume).

[7] See Johnson (2005, 64–80) and Henry (this volume).

thereby preserve themselves (III.12, 434b1–8). So "sense perception . . . is thus ultimately present for the sake of enabling locomotive animals to gather food after birth" (Leunissen 2010, 66). And the capacity of touch is not only a capacity to perceive tactile properties. Animals also exercise this capacity in a way that enables them to seize that which is advantageous and flee from that which is deleterious when they come into contact with these objects (434b10–18). This, too, helps animals to preserve themselves, especially since taste is a variety of touch (434b19–21). So "[t]he presence of the capacity of touch is equally explained teleologically as being ultimately necessary for the sake of the self-preservation of *each* kind of animal" (Leunissen 2010, 66). In each of these cases, the exercises of an organism's perceptual capacities benefit the organism insofar as it possesses nutritive capacities.

According to the capacity view of soul, if a higher capacity is for the sake of a lower capacity, this relation will not be contained in either capacity's account. But the teleological subordination of an organism's higher capacities to its lower capacities binds them together and effects a sort of unity. To perceive is to manifest a form of life that is essential to what it is to be an animal. But this is because perception is necessary if an animal is to live (or live a better) nutritive life. Perception is for the sake of nutrition in the sense that it enables or facilitates this basic life activity and ultimately enables an organism to (better) preserve itself as the kind of being it is.

Now surely it is correct that an organism's higher powers aid the activities of the lower (and vice versa). But life is said in many ways; the life of nutrition is but one among several ways life manifests itself (II.2, 413a20–26). Though I will not argue for this in detail until section four, it is important to note right away that when Aristotle says that to perceive is to live in a distinctive way he is not making the claim that to live a perceptual life is among the ways to live a nutritive life. Indeed, Aristotle objects to the univocal definition of life in terms of nutritive activity that Dionysius of Chalcedon proposes by invoking life's homonymy.[8] Aristotle is clear that in animals "there must be a single identical part in virtue of which it is alive and in virtue of which we call it an animal" (*de Iuv.* 467b21–23).

[8] [I]f, then, the definition applies in a like manner to the whole range of the homonym, it does not define any one of the objects described by the term. This is what happens to Dionysius' definition of life when stated as a movement of a creature sustained by nutriment, naturally present within it; for this is found in plants as much as in animals, whereas life does not seem to be predicated in virtue of a single form, but is one thing in animals and another in plants. (*Top.* VI.10, 148a25–30)

This is not the nutritive part of the soul. To live as an animal is to live a fundamentally perceptual life; it is not to live a nutritive life with the additional benefits that come from being able to perceive. Or so I will soon argue.

Still, there is something important in the suggestion that the soul's unity lies in something like teleological subordination. In order to view an organism's perceptual and rational activities as being for the sake of its nutritive activities, it is useful to think of the nutritive capacity in a way that goes beyond what the capacity view strictly allows. According to the capacity view, the account of nutrition will be like that of perception. Just as perception has a proper object, the perceptible, nutrition has a proper object, namely, food. But a complete characterization of the nutritive capacity includes more than a description of the way its exercises are directed upon its proper objects. The nutritive capacity is present for the sake of maintaining a living organism as the kind of being it is. For the nutritive capacity is a "capacity such as to maintain its possessor as such" (*DA* II.4, 416b18–19). And if all other vital capacities are present ultimately for the sake of nutrition, it is because they ultimately enable an organism to better preserve itself as the kind of being it is.

In the next section, I develop the view that activity for the sake of form is central to understanding the soul's unity. This account, however, does not employ any relation of teleological subordination. It is wrong, according to this second conception, to think of an animal's perceptual activity as ultimately being for the sake of its nutritive activity. Nevertheless, we can view an animal's perceptual activity as an activity that maintains its possessor as the sort of being it is, a natural unity whose form is a perceptual soul. The account of the soul's unity that this view leaves us with is, we will see, fundamentally different than the capacity view's account.

The soul as nature

According to the second conception of soul, when you gaze upon a living organism, what stands before you is not a multiplicity of capacities each performing an autonomous function. Nor is it a holistic complex built up from these basic activities. An organism's living is a single activity. A living organism is a form being realized.

One arrives at this view if one takes two claims to be central to Aristotle's account: (i) the soul is life's principle, and (ii) living organisms are natural unities. It follows from the conjunction of these two claims that explanations of vital activities must take a general form – these activities must be viewed as aspects or manifestations of a single, unfolding, end-directed activity. The end of this activity is an organism's form. To specify an organism's form is to characterize a mature or perfect specimen of the kind of organism in question. This form is prior to an organism's vital activities. That is, an organism's parts and tissues come to be as they are and act as they do for the sake of the organism's form, for the sake of bringing into being, as completely as can be, a perfect exemplar of a specific kind of life. I will show why these two claims and the method of inquiry they demand yield this conception of soul. This will require a brief discussion of what a nature is and what it is for the soul to be a nature.

A living organism's soul is its form (*DA* II.1, 412a19–20; *PA* I.1, 641a17–18; *Metaph.* Z 10, 1035b14–16), and while it is sometimes appropriate to speak of a natural entity as having both formal and material natures, it is, even in these contexts, nature as form that is fundamental. For "the form indeed is nature more than [or rather than, μᾶλλον] the matter."[9] So a living organism's soul is its form and this form is a nature.[10]

A nature is an internal principle of movement and rest. It is a natural being's form as well as the end for the sake of which its natural movements occur (II.7, 198a24–26). Everything that occurs by nature occurs for the sake of some end.[11] More precisely, it is natures themselves that Aristotle says are for the sake of an end (II.7, 198b4–5; II.8, 199a11–12). The formal and final causes of a natural body's movements are one because "the essence of a thing, i.e. the form for this [τὸ τί ἐστιν καὶ ἡ μορφή] is the end or that for the sake of which" (198b2–3; cf. II.8, 199a30–34). So if a nature is directed toward an end and this end, a form, is something's nature, then a nature is directed toward itself. For something to possess a nature is for that nature to be both the beginning and end of its natural movements. This is why it is appropriate to think of a nature as a "road toward nature (ὁδὸς εἰς φύσιν)" (II.1, 193b12–18). Natural movement is therefore a kind of formal perfection;

9 *Phys.* II.1, 193b6–7; cf. II.8, 198b8–19, 199a30–32; *PA* I.1, 640b29; *GC* II.9, 335b35–336a1; and; *Metaph.* Δ 4, 1015a13–15.

10 Though Aristotle never explicitly says that the soul is an organism's nature, it is overwhelmingly clear that he maintains this view. For the soul is life's principle (*DA* I.1, 402a6–7; II.2, 413b11–13; II.4, 415b8–14; *Metaph.* Δ 8, 1017b15–16). That is, the soul is an internal principle of movement and rest that, as with most things that possess natures (*Phys.* II.7, 198a25–26), is these movements' formal, final, and efficient cause (*DA* II.4, 415b8–11). The soul is the form of a natural body (II.1, 412b4–6) and its functions and movements are repeatedly referred to as being natural or as occurring in accordance with nature (II.4, 415a22–b3, 415b15–20, and III.5, 430a10–17). Moreover, the soul is, with some qualifications (*Metaph.* E 1, 1026a5–6 and *PA* I.1, 641a32–b10), to be studied by the student of nature (*DA* I.1, 402a4–7; 403a27–28; and *PA* I.1, 641a17–33).

11 Cf. *Phys.* II.5, 196b21; II.7, 198b4–5; II.8, 199a7–34, 199b9–26; *DA* III.12, 434a31–32; *PA* I.1, 641b11–642a1; and *Somn.* 2, 455b16–17.

it is movement that constitutes the (further) realization or perpetuation of something as the kind of being it is. The exercise of a natural capacity, when nothing interferes with it, results in that for the sake of which the exercise occurs, namely, the (further) realization or perpetuation of the nature qua form itself (II.8, 199b9–26; *PA* I.1, 641b23–642a1; *Metaph.* Z 7, 1032a23–25).

If an organism's form, its soul, is a nature, this analysis of natures should apply to them. And Aristotle is clear that it does. The soul is the internal principle and cause of those movements that are the manifestation of a living organism being what it is (*DA* II.4 415b12–14). Since "for living beings, to be is to live" (415b13), the soul is the principle of those activities that are the exercises of an organism's vital capacities. And as it is with most natural unities, the final, formal, and efficient causes of a living organism's vital movements coincide. For, "the soul is the cause and first principle of the living body . . . the soul is cause as being that from which the movement is itself derived, as that for the sake of which it occurs, and as the essence of bodies which are ensouled" (415b9–12). So insofar as a soul is a nature, one can't understand the soul's activities without understanding them as occurring for the sake of a form, namely, the soul itself.

This is especially clear if we restrict our focus to an organism's nutritive activities and consider what it is for them to have a soul as a natural principle, form, and end. Nutritive activities are activities of an organism's "ensouled body *qua* ensouled."[12] "The cause of nutrition and growth . . . is the soul," says Aristotle, "for in all things which are naturally constituted there is a limit and a proportion both for size and for growth; and these belong to soul, but not to fire, and to its account rather than to matter" (416a14–18). As we noted earlier, the activity of the nutritive soul qua threptic faculty is the exercise of a "capacity such as to maintain its possessor as such" (416b18–19). It performs this function insofar as it sustains the soul's continuous exemplification in the same individual organism. The activity of the nutritive soul qua reproductive faculty has the same function. However, given that terrestrial organisms are not eternal, it performs it insofar as it sustains the soul's continuous exemplification in a numerically distinct organism by duplicating the soul as best as it can in that organism.

[12] τὸ ἔμψυχον ἂν εἴη σῶμα τὸ τρεφόμενον, ᾗ ἔμψυχον (416b9–10). The soul is the cause and principle of the movements but the ensouled being is that which moves since it is the living organism that does what it does in virtue of (or with) its soul (ἄνθρωπον τῇ ψυχῇ, I.4, 408a34–b18).

So the movements that occur when an organism is engaged in nutritive activity, the movements whereby nutriment is processed and ultimately becomes living tissue, have a single, unitary nature as their principle and end. Nutriment and blood are moved in the order and ratio necessary for an organism to come to be or persist as such by a unitary nature and this nutritive activity is irreducibly for the sake of the organism's continuous formal exemplification.[13]

But it is not only an organism's nutritive movements that occur for the sake of the organism's form. *All* activities that arise from soul have an organism's form as their end. For Aristotle, the determination of an organ's function depends on the role the organ's coming to be serves in the coming to be of the whole organism to which it belongs (*PA* I.1, 640a33–b4). The capacity whose exercise constitutes proper functioning for an organ is a capacity whose existence is explained by the role it plays in the development of a unitary organism, where the latter development is itself due to an internal principle of movement and rest whose end is the exemplification of a form.[14] That is, an organ comes to be for the sake of φing only if the coming to be of the organ is part of the coming to be of an individual with a form that requires an organ that φs.

So, according to this second conception of soul, to perceive is not just to take on perceptible forms without matter. When one perceives, one is partially realizing the form for the sake of which one's capacity to perceive came to be in the first place. To perceive is to realize more completely the kind of being one is qua perceiver.

Reconciling the two conceptions

It is my contention that the second conception of the soul according to which it is a unitary nature is primary for Aristotle and can serve as the basis of a reconciliation of both conceptions. That is, if we take the second conception to be primary, we can better understand the method Aristotle employs in *de Anima* and can accommodate those aspects of the first conception that we must if we are to understand the complexity of individual organisms and the commonalities among organisms of different species. I will first tackle the program of *de Anima* and then outline what it is for a unitary soul to comprise various parts.

[13] I expand upon the details of these nutritive activities and the sense in which their principle is soul in Frey (2015).

[14] An influential interpretation along these lines is Gotthelf (1976).

The program of de Anima

It is easy to view *de Anima* in the following way.[15] Aristotle first provides an appropriate definition of the soul in *DA* II.1. The soul is the form, in the sense of first actuality, of a natural organic body that has life potentially.[16] Then, having properly defined the substance that is soul, Aristotle goes on to study the soul's attributes, that is, the soul's various capacities. The method he employs is the method appropriate to the study of capacities in general: one inquires into the proper objects of the activities that are the exercises of these capacities (II.4, 415a14–21).

But this is *not*, I contend, what Aristotle is doing. When Aristotle remarks that life is said in many ways, this affects the standing of the supposed definition he provides in *DA* II.1. He opens *DA* II.2 with the following complaint.

> Since it is from things which are obscure but more obvious that we arrive at that which is clear and more intelligible in respect of the principle involved, we must try again in this way to treat of the soul; for a defining statement should not only make clear the fact, as the majority of definitions do, but it should also contain and reveal the reason for it. (413a11–16)[17]

This criticism of many definitions is being levied here against his own attempt to define the soul in *DA* II.1. That is why, in the very next passage that proclaims the homonymy of life, Aristotle describes himself as "making a fresh beginning of our inquiry" (413a20).[18] For the fact that life is said in many ways shows that the definition of *DA* II.1 is just a schema. And no mere schema can capture the unity of a subject matter.[19] For if 'life' is said in many ways, this will affect any definition that contains 'life,' including Aristotle's definition of soul, viz. 'the first actuality of an organic body capable of life.' The soul is the first actuality of an organic body capable of feeding, *or* perceiving, *or* thinking. But what, if anything, unites these disjuncts?

15 (i) Interpretations roughly along these lines include Barnes (1971), Irwin (1988), Hicks (1907), Polansky (2007), and Johansen (2012). (ii) This section picks up on several themes in Rödl (unpublished) "Threptikon and Aisthetikon" and Ward (1996), though my positive picture differs significantly from that of each of these papers.

16 This is supposed to capture what is central to the three formulations of the definition Aristotle provides in II.1 at 412a19–21, 412a27–28, and 412b4–6.

17 Cf. *Post. An.* II.8, 93a4–5; II.10, 93b36; and *Top.* I.8, 103b8–12.

18 Aristotle also describes his original definition as one that provides only "a rough account and sketch of the soul in outline (τύπῳ μὲν οὖν ταύτῃ διωρίσθω καὶ ὑπογεγράφθω περὶ ψυχῆς)" (413a9–10). Cf. *Top.* I.1, 101a18–24; *HA* I.6, 491a7–14, and Bolton (1978, 259).

19 So I disagree with Barnes (1971, 103) when he claims that being disjunctive is no impediment to serving as an explanatory middle term.

Perhaps Aristotle's earlier attempt to define the soul is still informative. It puts forward the view that the souls are forms, that they are forms of natural bodies, and that these natural bodies are organic.[20] I will not here discuss whether any or all of these claims survive further scrutiny. But one thing is clear. Aristotle's initial attempt to define the soul does not provide the starting point of a science of life – it does not show that life, that of which the soul is principle, is a single subject.

So when the common man says that feeding is a way of living or that perceiving is a way of living, he doesn't yet have a proper grasp of what life is. As things stand at this point of the inquiry, 'life' may not designate a single thing at all. There may be a science of feeding, a science of perceiving, and a science of thinking. What is it about these three sorts of activity that makes them all cases of living rather than three distinct things that happen to share a single name?

We can see, now, why Aristotle adopts a sequential inquiry. When he investigates the nutritive, perceptual, and rational capacities in turn, he is not (only) attempting to understand the living or vital capacities. He is not just providing us with their accounts. He is attempting to understand why any of these capacities *are* living or vital capacities. He is trying to reveal the unity among them in virtue of which they are to be considered instances of living at all.

This fits well with Aristotle's earlier methodological remarks. Aristotle wonders how general an inquiry about soul will be.

> We must be careful not to ignore the question whether there is one definition [of the soul], as of animal, or whether there is a different one for each [species], as of horse, dog, man, and god, the universal, animal, being either nothing or secondary; and it would be similar for any other common predicate. (I.I, 402b5–9)

The scientific investigation of soul may begin with a definition that is as general as that of living beings. It may, however, begin with definitions as specific as horse, dog, man, and god. If the latter, there will be irreducibly different principles that govern horse life, dog life, human life, and divine life. We can call all of these things living; we can say that they all possess principles of life, viz. souls. But the words 'soul' and 'life' either (i) will pick out a collection of homonyms and provide no common understanding of why each of them receives these names – they will be nothing – or (ii) will

[20] On the suggestion that the definitions of soul in II.I are provisional, see Bolton (1978), Matthews (1992), and Ward (1996).

pick out a single thing but only after one has come to understand more specific instances – they will be secondary.

Aristotle rejects the idea that there can be a single, general definition of soul. Indeed, he says "it is foolish (γελοῖον) to seek both in these cases and in others for a common definition (κοινὸν λόγος) which will be a definition peculiar (ἴδιος λόγος) to no actually existing thing and will not correspond to the proper indivisible species (τὸ οἰκεῖον καὶ ἄτομον εἶδος), to the neglect of one which will" (II.3, 414b25–27). By Aristotle's own lights, his attempt to give a definition of soul in *DA* II.1 is foolish. So if there is to be a single science of soul, if 'soul' is not to designate nothing, then it must be secondary; it must be something that we can understand only after a thorough inquiry takes place at a lower level of generality.

What level of generality is appropriate, then, to an investigation of soul? We cannot begin at the level of dog, horse, man, and god. Even now, we haven't been able to survey the vast number of species given at this level of generality. And without such a survey, we could not discover what unifies all of these ways of being as instances of living. And we have already seen that a univocal definition of life that does not arise from an investigation at a lower level of generality is unavailable. But there is a level of generality that may work between that of life in its broadest use and the sundry species we encounter. Aristotle believes that if we study what feeds, then what perceives, then finally what thinks, this will reveal the unity of soul that will enable it to be the proper subject of a science.

What kind of unity will there be at this level of generality? Consider the following parallel problem. *Metaphysics* Γ declares that there is a science of being qua being. But to be (τὸ ὄν) is said in many ways (Γ 2, 1003a33). So, the question arises: How can there be a science of being qua being unless there is some unity to the various ways to say something *is*? Aristotle's answer employs what has come to be called 'focal meaning.' Substance is prior in account (λόγῳ, *Metaph.* Z 1, 1028a31–32) to the other ways in which a thing is said to be, and the unity this priority in account effects enables a scientific investigation into being qua being despite the irreducible multiplicity of ways in which 'to be' is said (Γ 2, 1003b12–18).

Life and soul face a similar challenge. But the route to achieve unity available in the case of being, namely, focal meaning, is not similarly available for life. One might think that the nutritive soul could serve as the focal relatum of the other forms of life.[21] For "the nutritive soul belongs

[21] Proponents of the focal meaning analysis include Owen (1960), Lloyd (1962), Leszl (1970), and Simplicius (1882). Shields (2002, 176–93) emphasizes Aristotle's claim that "the actuality of thought

also to the other living things and is the first and most commonly possessed potentiality of the soul, in virtue of which they all have life" (II.4, 415a23–25; cf. III.12, 434a22–26). But even if we ignore the difficulties that arise when this claim is extended to celestial/divine life, this suggestion fails because whatever ontological priority the nutritive soul possesses does not entail that it is prior in account. As we have already seen, neither the capacity to perceive nor the capacity to think has an account that must refer to the nutritive capacity. And without this, there is no focal unity.

Perhaps the ways life is said possess an analogical unity.[22] But this suggestion faces its own difficulties. For a first thing is related to a second by analogy when they are "related as a third thing is to a fourth" (*Metaph.* Δ 6, 1016b35). The analogy, if it is to be successful, must reveal an equality of proportions (either quantitative or qualitative) between the two pairs of relata (*EN* V.3, 1131a30–b4). But it is difficult to see what this proportional relation could be given how different the activities of the vital capacities are.[23]

What is left if the basic ways life is said are not focally or analogically related? Aristotle's answer is that they are hierarchically related (*DA* II.3, 414b28–415a13). I contend that being a member of a properly grounded hierarchy is a genuine source of unity that cannot be reduced to the unity that either focal meaning or analogy affords. If there is a science of the soul, there must be a principle of the sequence that begins with things that manifest nutritive activity, continues with things that manifest perceptual activity, and ends with things that manifest rational activity. Each of these activities must be investigated in turn. It is the principle that grounds this series that allows there to be a science of life and it is by coming to understand this hierarchy that one understands life itself. This is why the "reason they [sc. the capacities of soul] are so arranged in order of succession must be considered" (414b33–34).

This is not just an empirical question. It is not enough simply to say that when we investigate the living organisms around us, we find that the basic modes of life are ordered successively and hierarchically. We do find, when we investigate the terrestrial realm, that plants possess only nutritive capacities, that animals possess the capacities of plants as well as perceptual capacities, and that humans possess the capacities of plants and animals

is life" (*Metaph.* Λ 7, 1072b26–27) and suggests that god's rational soul, not the nutritive soul, serves as the focal relatum.

22 Proponents of the analogical analysis include Alexander (1887) and Rodier (1900).

23 Ward (1996, 126) expresses a similar pessimism. Ward, however, takes definition, focal meaning, and analogy to exhaust the available options.

plus the capacity to understand and think. But just as there is a difference between investigating vital capacities and trying to understand what it is for a capacity to be vital, there is a difference between investigating a hierarchy of capacities and trying to understand the principle by virtue of which the hierarchy is ordered in the way we find it to be. The very possibility of a science of life depends on a satisfactory answer to this last question.

So Corcilius and Gregoric are mistaken when they say "providing successive accounts for the four capacities amounts to a satisfactory account of the soul, an account which is explanatory of the souls that actually exist in the world, namely the souls of plants, animals (sensile and mobile), and humans" (Corcilius and Gregoric 2010, 105). This assumes that we know what soul is before the investigation begins. There are three or four kinds of soul and the reason we need to investigate them all, they think, is simply to be thorough. But the reason we must study these capacities in turn is to make them intelligible *as* capacities of soul. And this requires that we focus not on the capacities as if they were autonomous subjects of empirical inquiry that fall under the genus 'capacity of soul,' but rather on the principle that grounds the series and allows us to understand them all as forms *of living*.

The unity of soul

What, then, is the principle that grounds the hierarchy of souls and enables us to understand life? My answer to this question will not be complete, but it will provide reasons for thinking that the second conception of soul, the soul as unitary nature, is primary for Aristotle.

When Aristotle introduces the series of souls, he compares the ordering of souls to the ordering of figures (σχήματος). He says,

> it is clear, then, that it is in the same way as with figure that there will be one definition of soul; for in the former case there is no figure over and above (παρὰ) the triangle and the others which follow it in order, nor in the latter case is there soul over and above those mentioned. Even in the case of figures there could be produced a common definition (λόγος κοινός), which will fit all of them but which will not be peculiar (ἴδιος) to any one. Similarly too with the kinds of soul mentioned. (II.3, 414b20–24)

Aristotle then goes on to describe these series. He says,

> The circumstances with regard to soul are similar to the situation over figures; for in the case both of figures and of things which have soul that which is prior always exists potentially (δυνάμει) in what follows in order,

> e.g. the triangle in the quadrilateral on the one hand, and the nutritive soul in the perceptual on the other. (414b28–32)

So to understand the unity of life that is secondary to an understanding of what feeds, what perceives, and what thinks is in part to understand the way in which the nutritive soul can exist potentially in the perceptual soul and the way in which the perceptual soul can exist potentially in the rational soul. That is, to understand the unity of life as it manifests itself across species of varying complexity, we must understand how the souls that are the principles of these forms of life are a unity within an individual organism.

To explicate exhaustively what it is for one thing to exist potentially (to be *dunamei*) within another is a difficult task. Aristotle is clear that neither what it is to be something *dunamei* nor what it is to be *in* something *dunamei* can be defined (*Metaph.* Θ 6, 1048a35–36). The way figures possess this manner of being is but one among several. To understand this manner of being requires that we survey the various cases in which it arises and, as Aristotle puts it, "be content to grasp the analogy" (1048a36). I cannot here provide a complete survey, but I will discuss two cases and draw out the consequences they have for understanding the soul's unity: (i) what it is for a triangle to be *dunamei* in a quadrilateral, the case Aristotle mentions explicitly in *de Anima*, and (ii) what it is for a collection of simple bodies that undergo mixing to be *dunamei* within the mixture that results.

Two features of the series of figures are relevant to the series of souls. First, the series is ordered by relations of priority and posteriority. The triangle is present *dunamei* in the quadrilateral and not vice versa. Second, to be in something *dunamei* is to exclude the other manner in which something can be. The triangle is not in the quadrilateral actually (ἐνεργείᾳ).

The first claim, that souls, like figures, form an ordered series (ἐφεξῆς), highlights a consequence for any class whose members are ordered by relations of priority and posteriority. Namely, for any ordered series, there is no genus prior to the species that are so related.[24] And if there is no prior,

[24] Cf. *Metaph.* B 3, 999a6–14; *Polit.* III.1, 1275a34–38, and especially *EE* I.8, 1218a1–8. Aristotle's reasoning is as follows. If there were a predicate common to all members of an ordered series, it would be prior to the members of the series (*EE* I.8, 1218a4). This priority is cashed out in terms of asymmetrical existential dependence: if what the common predicate picks out were not, the members of the series would not be, though not vice versa. But an ordered series' first member is prior to its subsequent members in the same sense (*Cat.* 12, 14a29–35; *Metaph.* M 8, 1083b32–34). If the first member were not, the subsequent members of the series would not be, though not vice versa. Further, a genus can be even if any one of the species that fall under it is not (*Cat.* 13, 15a4–7; *Top.* VI.4, 141b28–29; *Metaph.* K 1, 1059b38–1060a1), though a genus cannot be if all of the species that fall under it are not (*Cat.* 5, 2a35–b6). These commitments lead to a *reductio* of the assumption

common genus, then a science of soul is possible only if there is a single way to understand what it is for a prior member of the series to be present *dunamei* in a posterior member of the series.

The second claim about the series of figures that is relevant to the series of souls follows from the fact that being something *dunamei* and being something *energeiai* are mutually exclusive manners of being.[25] So if a prior figure (or soul) is present *dunamei* in a posterior figure (or soul), then the prior figure (or soul) is not what it is *energeiai*.[26] Though a quadrilateral has three angles and three sides (if it has four, then a fortiori it also has three), it is nevertheless a mistake to say that in addition to being a triangle, a quadrilateral is *also* a quadrilateral. Similarly, it is a mistake to say that in addition to leading a nutritive life, an animal *also* leads a perceptual life. Strictly speaking, there is no nutritive soul in a human or animal. At least it is not present in the way that the capacity conception suggests; it is not present *energeiai*. There is, in an animal, *only* a perceptual soul. The nutritive soul is present *dunamei*. To say that an animal not only perceives but *also* feeds suggests that perceiving is something the animal does *in addition to* feeding. But this is wrong. For an organism, to be is to live, and for an animal, to live is to perceive. So perceiving is not simply something that an animal does. An animal perceives and *that is all*. There is, of course, a secondary sense in which animals feed. It would be a mistake not to attribute nutritive *capacities* to animals. But when an animal exercises its nutritive capacity, this activity does not have a nutritive soul as its principle. *The principle of an animal's nutritive activities is a perceptual*

that the species in an ordered series fall under a common genus. For if there were such a genus, (i) the genus would be even if the first species in the series were not, (ii) if the first species were not, none of the species in the series would be, and (iii) if none of the species are, the genus would not be. Caston (unpublished) "Aristotle on the Unity of Psychology: How to Divide the Soul," contains a clear and thorough discussion of these arguments.

25 Cf. *Phys.* III.1, 201a20–23, 201b6–15; *Metaph.* Θ 6, 1048a30; and Frey (2015).

26 Johansen explicitly denies that this is true of the series of souls:

> saying that the nutritive soul is potentially present within a living being with perception may suggest that it is not actually present. But this cannot be right: animals eat, drink, and procreate, so their nutritive soul must be present in activity . . . the claim is not that the nutritive soul as such is not actually present in an animal; the point is rather that it is implied in saying that the living being has perception that it also has nutrition. (2012, 69)

The impending discussion supports the claim that the nutritive soul is not present *energeiai* in animals. But one thing to note right away is that Johansen's view would serve as a serious impediment to the soul being a unity since "a substance cannot consist of substances present in it in actuality, for things that are thus actually two are never actually one, though if they are potentially two, they can be one" (*Metaph.* Z 13, 1039a4–6). As we will see, an animal can have nutritive capacities – it can eat, drink, and procreate – without this implying that the animal possesses a nutritive soul *energeiai*.

soul. Everything an animal does has its one and only soul as its principle and end. So feeding is one of the ways in which the animal's single soul, a perceptual soul, is actualized.

So if one provides accounts for the various capacities that living organisms exercise and also gives a complete description of the numerous external relationships of teleological subordination in which these capacities stand to one another, one has not yet understood the soul's unity. Doing these things does not equip one with an understanding of what it is for a lower soul to be present *dunamei* within a higher soul. To understand what it is for a nutritive soul to be present *dunamei* in a perceptual soul is to understand how a perceptual form of life can manifest itself through feeding, and to understand what it is for a perceptual soul to be present *dunamei* in a rational soul, is to understand how a life of thought can manifest itself through both feeding and perceiving. It is only when we understand what it is for each of these souls to be united in this way within an individual that we come to understand the hierarchical ordering of the basic ways life is said; it is only then that we come to understand life itself.

What is it, then, for a lower soul to be present *dunamei* in a higher soul and what is it about this presence that enables the higher soul to be the principle of those vital activities that would have the lower soul as their principle if the lower soul were *energeiai*? It is comparatively easy to see what it is for a triangle to be *dunamei* within a quadrilateral. Aristotle says that "in potentiality the half line is prior to the whole line . . . but in actuality it is posterior; for it is only when the whole is dissolved that it will exist in actuality" (*Metaph.* Δ 11, 1019a8–11; cf. Θ 6, 1048a32–33). What Aristotle says about lines holds for figures as well. That is, neither lines nor complex figures are composed of actual half lines or actual triangles. They are not aggregates of simpler elements (Z 17, 1041b11–33). But a whole line can be divided into halves, and, when it is, the half lines transition from being what they are *dunamei* to being what they are *energeiai.* Similarly, a quadrilateral can be divided into two triangles, and, when it is, the triangles transition from being what they are *dunamei* to being what they are *energeiai.*[27]

But this can't be a satisfactory account of the manner in which the nutritive soul is present *dunamei* in the perceptual soul. There is no means by

[27] This separation need not consist in a single act of division. One can separate triangles from quadrilaterals in this way, but not, say, quadrilaterals from pentagons. But more complex figures contain similar units, namely sides, from which one can construct simpler figures and, in this way, the simpler figures can be separated out from the more complex.

which an animal's single soul can be divided in such a way that the nutritive soul present in it *dunamei* comes to be what it is *energeiai*. Though there is a practice of calling the state of those with seriously impaired rational and perceptual capacities 'vegetative,' the soul, according to Aristotle, would not be changed under these conditions. We ought not to describe such individuals as possessing only a nutritive soul. In fact, Aristotle is clear that these individuals still possess perceptual and rational capacities. The souls of these individuals are exactly the same as they were when they were healthy. Only the organs required to exercise these capacities have been affected and if the bodily damage were fixed, the capacities that the organism possesses before it is healed could be exercised once again (I.4, 408b21–24). And though Aristotle thinks that an organism gains its vital capacities sequentially – a developing animal embryo does not possess the capacity to perceive until the organs required for perception come to be (*GA* II.3, 736b21–26 and IV.1, 766a5–10) – it does not follow that an embryo first possesses a nutritive soul, and then only later possesses a perceptual soul. The soul that is an animal embryo's principle of life is, from the beginning, a perceptual soul. That is why animal embryos in their early developmental stages, unlike plants that only have a nutritive soul, are such that they can come to possess perceptual capacities *energeiai* (*GA* II.3, 736b13–15).

So the series of figures and the series of souls are not similar in all respects. This is to be expected given that the feature of these series that concerns us most, what it is for their prior members to be *dunamei* in their posterior members, can only be understood analogically. We can make headway, however, in understanding the way this relation applies to souls if we turn to our second analogical case, namely, mixtures (μικτόν).[28]

A mixture is a homoeomerous natural body and therefore has a single, unitary nature. But this nature is not primitive. That is, when a mixture comes to be, none of the simple ingredients from which it comes to be cease to be; mixing is neither growth nor generation (*GC* I.10, 327b3–8). And when a mixture comes to be, the ingredients from which it comes to be cannot exist in the same way they did before they were mixed; a mixture is not an aggregate and mixing is not alteration (327a35–b3). But Aristotle has the resources to accommodate both the continued presence of the ingredients in a mixture and the fact that what it is for these ingredients to be when they are present in a mixture must differ from the way they are before they are mixed. Aristotle says that, "it is possible for things after they

[28] For a more thorough account of mixture, see Frey (2007, 187–90 and 194–97).

have been mixed to be and not to be. Some other thing which comes to be from them is actually (ἐνεργείᾳ), while each of the things which were, before they were mixed, still is, but potentially (δυνάμει), and has not been destroyed" (327b23–26). So the ingredients from which a mixture comes to be are present in it *dunamei.*

That a mixture's ingredients are present in it *dunamei* is not an idle claim. We can't understand a mixture's nature if we don't appreciate the role this presence plays in a mixture being what it is. A mixture's unitary nature is the principle of all of its tangible movements. These movements are intermediate with respect to the natural movements of the ingredients that are present in it *dunamei* (II.7, 334b9–16). There is a continuum that has maximally hot movements and maximally cold movements as extremes. If a mixture comes to be from one ingredient that is hot, for example fire, and one that is cold, for example water, it will have a single thermal capacity whose exercise results in movements that are located somewhere in the middle of this continuum.[29] This capacity is not primitive. It is not a new capacity – warm – that has the same status as fire's heat or water's coldness. The mixture's capacity is not only aetiologically dependent on the capacities of its ingredients, but ontologically dependent too. That is, the mixture could not possess its natural capacity, hot-cold, if the ingredients from which it came to be were not present in it *dunamei.* To possess this capacity just is for these ingredients to be present in this manner of being. So mixtures have a single, unitary principle of movement and rest, but this principle is a principle of movements that it could not undergo without the ingredients from which it came to be being present in it *dunamei.*

Given that living organisms are also natural unities, we can draw several parallels between the natures of mixtures and souls. (i) Just as a mixture's ingredients are not present in it *energeiai*, an animal's nutritive soul is not present in it *energeiai.* (ii) Just as the movements of mixtures have a unitary nature as their principle and end, a living organism's vital activities have a unitary nature as their principle and end, namely, a soul. (iii) Just as a mixture could neither possess the tangible capacities it does nor have a nature that is the principle of its tangible movements unless its ingredients were present in it *dunamei*, an animal could neither possess the nutritive capacities it does nor could its perceptual soul be the principle of its nutritive activities unless a nutritive soul were present *dunamei* in the

29 There are three continua that capture the capacities natural bodies possess insofar as they are tangible. The extremes of these continua are fixed by three pairs of contraries: hot/cold, wet/dry, and heavy/light. What is said here about thermal capacities applies to the hygric and locomotive capacities as well.

animal's perceptual soul. And (iv) just as the natural unity of mixtures neither prevents them from possessing various capacities for movement nor prevents these capacities from having definitions that are separable in account in such a way that bodies with different natures can possess them, the natural unity of living organisms neither prevents them from possessing various vital capacities nor prevents some of these capacities from having definitions that are separable in account in such a way that organisms of different species can possess them.

Again, as should be expected in any analogical enterprise, the case of souls does not map onto that of mixtures perfectly. For example, the natural movements of the simple bodies and mixtures are located on continua. Moreover, there is an effective rule of composition that one can employ to derive a mixture's intermediary capacities from those of its ingredients. Neither of these is true for the basic ways life is said. There is no continuum on which to place both nutritive and perceptual activities. And there is no simple rule to which one can appeal in order to understand the principle of their hierarchical ordering. Without this understanding, we have not fully understood what life is. So the present study is, in an important sense, incomplete. Indeed, to provide a complete explanation of what it is for a prior soul to be present *dunamei* in a posterior soul is, I contend, among the principal tasks of *de Anima* as a whole. But what has been put forward already suggests that whatever answer is forthcoming, it will not be consistent with the capacity conception of the soul. What we need is a notion of soul that sees its activities as arising from a single, unitary principle of movement and rest and as occurring for the sake of a single, formal end. This is exactly what the conception of soul as nature provides.

PART III

Methodology

CHAPTER 9

Aristotle's architectonic sciences

Monte Ransome Johnson

Aristotle rejected the idea of a single, overarching super-science or "theory of everything," and he presented a powerful and influential critique of scientific unity. In theory, each science observes the facts unique to its domain, and explains these by means of its own proper principles. But even as he elaborates a prohibition on kind-crossing explanations (*Posterior Analytics* 1.6–13), Aristotle points out that there are important exceptions – that some sciences are "under" others in that they depend for their explanations on the principles of a superior (more architectonic) science. In this chapter, I explore how subordination relations and architectonic structures apply to Aristotle's scientific practice – including not only the works of theoretical philosophy, which have already been discussed in this connection,[1] but also between these and the practical and productive sciences.

After reviewing essential Platonic background in part I, in part II I ask: What is the "most architectonic" science for Aristotle? Aristotle appears to give conflicting answers in the opening paragraphs of the *Metaphysics* and the *Nicomachean Ethics*. The apparent tension can be resolved by distinguishing two ways for sciences to be subordinate: (i) with respect to their ends, where one science exists for the sake of another; and (ii) with respect to their objects, where one science treats a subject matter at a higher level of generality than another, related science. In part III, I examine the ways in which Aristotle divides philosophy into the sciences and conceives of their interrelationship. Aristotle considers all productive sciences to be

I'd like to acknowledge written comments I received from D. S. Hutchinson and David Ebrey on an early draft of this chapter that considerably helped me to improve the version I later presented at UCLA in February 2013, where I received further important comments and criticisms from Henry Mendel, Gavin Lawrence, Robert Bolton, David Blank, Alan Code, Richard McKirahan, Klaus Corcilius, and Sean Kelsey. A still further revised draft was discussed at the UCSD History of Philosophy Roundtable; I'd like to thank my colleagues Eric Watkins, Sam Rickless, and Michael Hardimon, as well as Blythe Greene and Ashley Attwood, for their valuable constructive criticism. I'd also like to thank Julene Knox for helpful editorial advice.

[1] See: McKirahan 1978, Lennox 1986, and Hankinson 2005.

subordinate to practical sciences on the basis of their ends (and thus the practical sciences are architectonic over them), but in turn all practical sciences are subordinate to theoretical sciences. The theoretical sciences are superior to the others on the basis of both ends and the generality of objects, being at once the most valuable, since they are intrinsic goods worthy of being pursued for their own sake; and at the same time the most general, concerned with the ultimate causes "for the sake of which." Thus the theoretical are, in the final analysis, the "most architectonic" sciences.

In part IV, I develop an experimental model of the architectonic structure of Aristotle's theoretical sciences in syllogistic form. In part V, I apply this model to show how theoretical and in particular natural science might function as architectonic over political science.

I

In the *Politicus*, Plato distinguishes between arithmetic and other "arts closely related to it stripped of practical activities, but that provide only thinking" (τέχναι ψιλαὶ τῶν πράξεών εἰσι, τὸ δὲ γνῶναι παρέσχοντο μόνον, 258d4–6) and "the kinds of art involved in carpentry and manufacture as a whole, that have their knowledge such that it is naturally grafted to practical actions, and that bring to completion those corporeal things that previously did not exist" (258d8–e2); and so he asks us to "divide all of science in this way, designating one kind the *practical*, but another kind the purely *intellectual*" (ταύτῃ τοίνυν συμπάσας ἐπιστήμας διαίρει, τὴν μὲν πρακτικὴν προσειπών, τὴν δὲ μόνον γνωστικήν, 258e4–5). The resulting division seems prima facie to anticipate Aristotle's distinction between "practical" and "theoretical" sciences, and so Rowe translates γνωστικήν as "theoretical" here (1997). But Aristotle's change of terminology is unlikely to be insignificant, and so I follow Fowler's translation "intellectual" (1925) in order to keep the term "theoretical" for Aristotle's θεορετική.

In a nearby passage, Plato articulates in rapid dialectical succession two theses, one that Aristotle vehemently rejects, and another that he emphatically embraces. So Aristotle opens *Politics* I with a rejection of the claim of the *Politicus* that: "surely there won't be any difference, as far as ruling is concerned, between the character of a large household, on the one hand, and the bulk of a small city, on the other" (259b9–10, tr. Rowe) and that, in consequence "there is one kind of expert knowledge concerned with all these things, whether one gives the name of expertise in kingship, or statesmanship, or household management" (259c1–4, tr. Rowe, adapted; contradicted by Aristotle at *Pol.* I.1.1252a7–16). We will return to that argument in due course. I want now to point out that Aristotle agrees with the immediately following thesis: "this much is clear, that the power of any king to maintain his rule has little to do with the use of his hands or his body in general, in comparison with the understanding and force of his mind" (259c6–8, tr. Rowe), so that "the king is more closely related to *the intellectual kind of knowledge* than to the manual or wholly *practical kind*" (259c10–d2, tr. Rowe, adapted).

Now it is striking to see politics classified not as a *practical* science but as an "intellectual" (what Rowe calls "theoretical") science, especially for those students of Aristotle taught to sharply distinguish between these, where politics is invariably classed with the practical, not theoretical, sciences. But there is a very important sense in which Aristotle agrees with Plato, however, and realizing that must cause us to revise the conception of political science as simply and straightforwardly a practical science.

Plato associates practical science with the activities of manual laborers, and it is against this class that he opposes his "purely intellectual" sciences, which he instances with arithmetic and the art of calculation (259e5), but also with the activities of "all architects (ἀρχιτέκτων) too – since they don't act as workers themselves, but manage workers" (259e9–10); thus the architect "provides some understanding rather than manual labor" (259e11). But these two sciences, calculation and architecture, are distinguished insofar as "one makes judgments, while the other directs." These sciences, then, are the archetypes of the highest division of "purely intellectual" science: "referring to one as directive and the other as making judgments" (260b3–5, tr. Rowe, adapted). Let us outline these two divisions of the arts and sciences as follows:

(1) The practical (τὴν μὲν πρακτικήν), i.e. manual labor
(2) The purely intellectual (τὴν δὲ μόνον γνωστικήν):
  a. The directive (τὸ μὲν ἐπιτακτικὸν μέρος), e.g. architecture, statesmanship
  b. The critical (τὸ δὲ κριτικόν), e.g. arithmetic, calculation

But instead of a division, we should arrange these sciences into a hierarchy:

Intellectual science:	1. Architecture (e.g. directing the building of a temple)
Practical science:	2. Manual labor (e.g. creating bricks, moving bricks)

Such an arrangement then suggests the possibility that another "intellectual" science is somehow superordinate over architecture:

Intellectual sciences (critical):	1. Arithmetic, calculation, etc.
Intellectual sciences (directive):	2. Architecture, statesmanship, etc.
Practical sciences:	3. Manual labor

Such an "architectonic" arrangement of sciences, in effect, is embraced by Aristotle, and extended to several theoretical and practical sciences, or so I will argue in what follows.

II

Aristotle opens the *Nicomachean Ethics* by describing the subordination of certain arts and sciences on the basis of their ends:

> As there are many actions, arts, and sciences, their ends too are many; the end of the medical art is health, that of ship-building a vessel, that of strategy victory, that of economics wealth. But where such arts fall under a single capacity – as bridle-making and the other arts concerned with the equipment of horses fall under the art of riding, and this and every military action under strategy, in the same way other arts fall under yet others (κατὰ τὸν αὐτὸν δὴ τρόπον ἄλλαι ὑφ' ἑτέρας) – in all of these the ends of the architectonic arts are to be preferred to all the subordinate ends (ἐν ἁπάσαις δὲ τὰ τῶν ἀρχιτεκτονικῶν τέλη πάντων ἐστὶν αἱρετώτερα τῶν ὑπ' αὐτά); for it is for the sake of the former that the latter are pursued. It makes no difference whether the activities themselves are the ends of the actions, or something else apart from the activities, as in the case of the sciences just mentioned. (*EN* 1.1, 1094a6–18, tr. Ross, adapted; by Water's text)

All of the examples in these passages are of practical and productive sciences. Let us outline a hierarchy by means of one of the examples:

(Architectonic) Practical science:	1. Strategy
(Subordinate) Practical science:	2. Art of horse-riding
(Subordinate) Productive science:	3. Art of bridle-making

Science 3 is for the sake of science 2, which is in turn for the sake of 1 – a straightforward hierarchy based on the ends of the sciences. The bridle makers perform their activities for the sake of the horse riders (i.e. the cavalry) and not vice versa. The productive science is directed by a superior practical science, but one that is in turn subordinate to a still superior practical science: strategy. Both the content of the examples (strategy, cavalry, ship-building, etc.), and also the logical structure of the ranking, suggest the kind of division of labor and hierarchy at home in military logistics.

Aristotle introduces the *Nicomachean Ethics* this way because he is building up to the question of what is the "most architectonic" science, which will have as its proper object the "chief" good:

> We must try, in outline at least, to determine what it is, and of which of the sciences or capacities it is the object. It would seem to belong to the most authoritative art and that which is most architectonic (δόξειε δ' ἂν τῆς κυριωτάτης καὶ μάλιστα ἀρχιτεκτονικῆς). And politics appears to be this kind of thing (τοιαύτη δ' ἡ πολιτικὴ φαίνεται); for this ordains which of the sciences should be studied in a state, and which each class of citizens should learn and up to what point they should learn them; and we see even the most highly esteemed of capacities to fall under this, e.g. strategy, economics, rhetoric; now, since politics uses the rest of the practical[2] sciences (χρωμένης δὲ ταύτης ταῖς λοιπαῖς πρακτικαῖς τῶν ἐπιστημῶν), and since, again, it legislates as to what we are to do and what we are to abstain from, the end of this science must include those of the others, so that this end must be the good for man. (*EN* 1.2, 1094a25–b7, tr. Ross, adapted)

Let us represent the kind of hierarchy described in the last two passages as follows:

Practical science:	1. Politics
Practical science:	2. Strategy
Practical science:	3. Art of horse-riding
Productive science:	4. Art of bridle-making

In this scheme, strategy is one of the highest-level divisions of a science over which politics is architectonic. Strategy, although subordinate to politics, is in its rank architectonic over the command of the cavalry and the associated productive sciences (e.g. bridle-making). On the same highest-level division as strategy would be economics, rhetoric, etc. Under each of these divisions are further divisions, and under those still further divisions and relations of subordination. Aristotle presupposes at least this level of complexity of architectonic sciences in the opening words of the *Nicomachean Ethics*.

2 Reading πρακτικαῖς at 1094b4 (found in all MSS), deleted by Bywater (1984); Ross follows this in the Oxford translation (1925), as does Urmson in the Revised Oxford Translation (Ross 1984). It is also omitted (without explanation) from the translations of Rackham (1926) and Crisp (2000), but retained, rightly, in the translations of Irwin (1999) and Rowe (2002). Gauthier and Jolif in their commentary on 1094b4 are supportive of the deletion on the grounds that it appears to be an imprecise gloss, since it fails to mention Aristotle's productive sciences. But as I will argue below, Aristotle himself frequently resorts to a dichotomy between "practical" and "theoretical," using the term "practical" (and sometimes "productive") to include both the practical and productive sciences as a group.

Can this passage be read as evidence that Aristotle considered politics architectonic over all other arts and sciences without qualification? The reason he says politics appears to be so – because politics ordains which other sciences are to be studied and up to what point – would seem to support such a reading. But it is difficult to imagine that Aristotle would accept the politician directing the natural scientist or metaphysician in the manner that an architect would direct a manual laborer. And if we turn to the parallel text in *Metaph.* 1.2, we find Aristotle most definitely changing his tune:[3]

> Of the sciences, that which is more desirable for its own sake and thanks to the knowing (τὴν αὑτῆς ἕνεκεν καὶ τοῦ εἰδέναι χάριν αἱρετὴν) is more wisdom than that which is desirable for the sake of its results, and the more over-arching science is more wisdom than the subordinate science (τὴν ἀρχικωτέραν τῆς ὑπηρετούσης μᾶλλον εἶναι σοφίαν); for the wise man must not be directed but must direct (ἐπιτάττειν), and he must not obey another, but the less wise must obey him . . . The first principles and causes (τὰ πρῶτα καὶ τὰ αἴτια) are the most knowable; for by reason of these, and from these, all other things are known, but these are not known by the things underlying them (διὰ γὰρ ταῦτα καὶ ἐκ τούτων τἄλλα γνωρίζεται ἀλλ' οὐ ταῦτα διὰ τῶν ὑποκειμένων). And the science which knows that for the sake of which each thing is done is the most over-arching of the sciences (ἀρχικωτάτη δὲ τῶν ἐπιστημῶν) and is more over-arching than any subordinate one (μᾶλλον ἀρχικὴ τῆς ὑπηρετούσης) and this end is the good in each class and in general the supreme good in the whole of nature. Judged by all the tests that we have mentioned, then, the name in question falls to the same science; this must be a science that investigates the first principles and causes; for the good, i.e. that for the sake of which, is one of the causes. That it is not a *productive* science is clear even from the history of the earliest philosophers. (*Metaph.* 1.2, 982a14–19+982b2–12, tr. Ross, adapted)

EN 1.2 and *Metaph.* 1.2 parallel each other not only in being placed at the beginning of their respective inquiries, but also in the notable use of superlatives ("most architectonic"; "most over-arching") that indicate a single scientific hierarchy, and not a manifold of independent and non-overlapping

[3] Alan Code has stressed to me the importance of taking into account the different rhetorical situations of the *Ethics* and the *Metaphysics* and (as we will see later), of the *Protrepticus*. This is an extremely important point that I wish I had the space to comment on more extensively. Here I will only say that while the difference in audience is certainly a factor, the exact rhetorical genre of *EN* 1.1–2, *Metaph.* 1.1–2, and the *Protrepticus* is the same in each case: exhortation (discussed and defined by Aristotle at *Rhet.* 1.4–7). This goes a long way toward accounting for the similarities (e.g. why comparative and superlative terms abound in discussing the status of the activity in question). But this point about the rhetorical genre does not in itself address the differences in intended audience that Code mentioned, but only confirms it is important to take them into consideration.

scientific domains. And thus there seems to be a tension between the suggestions in the *Ethics* that the "most architectonic" science is politics, but in the *Metaphysics* that the "most over-arching" science is metaphysics (i.e. "the science which investigates first principles and causes"). After all, Aristotle elsewhere makes it clear that politics and wisdom are not the same thing:

> It would be absurd to think that the art of politics, or intelligence (τὴν πολιτικὴν ἢ τὴν φρόνησιν), is the most important thing (σπουδαιοτάτην), since man is not the best thing in the world . . . It is evident also that wisdom and the art of politics cannot be the same (οὐκ ἂν εἴη ἡ σοφία καὶ ἡ πολιτικὴ ἡ αὐτή); for if the state of mind concerned with a man's own interests is to be called wisdom, there will be many wisdoms; there will not be one concerned with the good of all animals (any more than there is one art of medicine for all existing things), but a different wisdom about the good of each species. (*EN* 6.7 = *EE* 5.7, 1141a20–33, tr. Ross, adapted)

And back in the *Metaphysics* he refers to this discussion in the *Ethics* as if it had resolved the issue of the relative merits of the different kinds of science:

> We have said in the *Ethics* what the difference is between art and science and the other kindred faculties; but the point of present discussion is this, that all men suppose what is called wisdom to deal with the first causes and principles of things. This is why, as has been said before, the man of experience is thought to be wiser than the possessors of any perception whatever, the artist wiser than the men of experience, the architect than the manual laborer (χειροτέχνου δὲ ἀρχιτέκτων), and the *theoretical* sciences to be more of the nature of wisdom than the *productive* sciences. Clearly then wisdom is knowledge about certain causes and principles. (*Metaph.* 1.1, 981b25–982a3, tr. Ross)

Notice that Aristotle utilizes the Platonic distinction (in the *Politicus*) between directive "architectonic" arts on the one hand, and practical "manual" arts on the other, in order to rank-order and at the same time motivate the study of theoretical science. Aristotle in effect subsumes Plato's hierarchy in his own rank-ordering of sciences:

(1) Theoretical sciences (science of first principles and causes, natural science, etc.)
(2) Productive science:
  a. Architectonic (politics, strategy, economics, etc.)
  b. Manual (bridle-making, etc.)

The key to resolving the apparent tension between the *EN* 1.2 passage and *Metaph.* 1.2 on the issue of the "most architectonic" science is to emphasize the "appears" (φαίνεται) in the statement "politics *appears* to be this kind of thing" (1094a27–28). The reality, it becomes clear, is quite different. The fact that politicians direct which sciences are to be studied in the state and to what extent makes it appear that politics is the most architectonic science, but the truth is that politics, although the most architectonic practical science, is itself subordinate to theoretical science, and in at least two ways. First, because the "science of first principles and causes" investigates "the good in each class and the supreme good in nature," and so it investigates the more general cause of which the political good is either a part or a species. Second, because theoretical science, unlike the practical and productive sciences, is an end in itself and desirable for its own sake, so that political science must ultimately exist for the sake of producing theoretical science. For both reasons, then, philosophy is the "most directive," "most commanding," and "most architectonic" science. Such a view is clearly expressed in a passage of the *Protrepticus* wherein Aristotle describes the productive sciences as in general subordinate to the practical,[4] and asserts that philosophy is the master science of them all:

> There is a difference between the kinds of science that produce (αἱ ποιοῦσαι) each of the things of which we want to have more and more in our way of life, and the kinds of science that make use of these (αἱ χρώμεναι ταύταις); and the ones that are subordinate (αἱ ὑπηρετοῦσαι) are different from the others that direct (αἱ ἐπιτάττουσαι); and in these as it were more commanding kinds (ἡγεμονικωτέραις) exists what is good in the strict sense. If, then, only that kind of science which does have correctness of judgment, and does use reason, and observes the good as a whole – that is to say, philosophy – is capable of using everything and issuing orders in accordance with nature,

[4] Here are two other passages that make it clear that productive sciences are in general subordinate to the practical:

> The arts which govern the matter and have knowledge are two, namely the art which uses the product, and the art which directs the production of it (αἱ ἄρχουσαι τῆς ὕλης καὶ αἱ γνωρίζουσαι τέχναι, ἥ τε χρωμένη καὶ τῆς ποιητικῆς ἡ ἀρχιτεκτονική). That is why the using art is in a sense directive; but it differs in that it knows the form, whereas the art which is directive as being concerned with production knows the matter. For the helmsman knows and prescribes what kind of form a helm should have, the other from what wood it should be made and by means of what operations. (*Phys.* 2.2, 194a36–b7, tr. Hardie and Gaye, adapted)

> Intellect itself, however, moves nothing, but only the intellect which aims at an end and is practical; for this rules the productive intellect as well (διάνοια δ' αὐτὴ οὐθὲν κινεῖ, ἀλλ' ἡ ἕνεκά του καὶ πρακτική· αὕτη γὰρ καὶ τῆς ποιητικῆς ἄρχει). For everyone who produces does so for the sake of something, and the product is not the end absolutely, but only relative to something; for the absolute end is what is to be done (ἀλλὰ τὸ πρακτόν), because acting well is the end, and desire is for this. (*EN* 6.2 = *EE* 5.2, 1139a35–b4)

> by all means one ought to do philosophy, since only philosophy includes within itself this correct judgment and this intelligence to issue orders without errors. (Aristotle, *Protrepticus*, apud Iamblichus, *Protr.* 6.37.11–22, tr. Hutchinson and Johnson, adapted)

Aristotle has superimposed on the Platonic distinction between "ordering sciences" (αἱ ἐπιτάττουσαι) and "subservient sciences" (αἱ ὑπηρετοῦσαι) his own distinction between "productive sciences" (αἱ ποιοῦσαι) and the practical sciences, which are here called "those that use the productive sciences (αἱ χρώμεναι ταύταις)."[5] In discussing the distinction between art and action in the *Ethics*, Aristotle may refer to this passage.[6] Whereas Plato had identified the practical and productive sciences (or rather, failed to distinguish them) and treated them as subordinate to "intellectual" science, Aristotle divides off the sciences that *produce* material goods from those that *use* them in order to bring about the good life, calling the productive subservient to the practical. Like Plato in the *Statesman*, however, Aristotle in the *Protrepticus* blurs the distinction between the theoretical and practical sciences by arguing that the ultimate "directive" and "most commanding" art is not a practical science but a theoretical one, which he repeatedly calls philosophy.[7]

[5] Recall that in *EN* 1.2 Aristotle asserts that "politics uses the rest of the practical sciences" (1094b4–5, according to the reading of the MSS, see above on Note 2).

[6]
> Among things that can be otherwise are included both things produced and things done; producing and acting are different, *we are convinced about these things even in the exoteric works;* so that the reasoned state to act is different from the reasoned state to produce. Nor are they included one in the other; for neither is acting producing, nor is producing acting. Now since building is an art and is essentially a reasoned state to produce, and there is neither any art that is not such a state nor any such state that is not an art, art is identical with a state to produce involving a true course of reasoning. (*EN* 6.4 = *EE* 5.4, 1140a1–10, tr. Ross, adapted and emphasis added)

[7] This conceptual scheme of despotism and leadership as applied to psychological and epistemological issues is common not only in Plato but also in Isocrates, who wrote in the *Antidosis*:

> Our nature is compounded of two parts, the bodily and that of the soul, and no one would deny that of these two the soul is more of a natural leader (ἡγεμονικωτέραν πεφυκέναι τὴν ψυχὴν) and is of greater worth; for it is the function of the soul to deliberate both on personal and on public questions, and of the body to be servant (ὑπηρετῆσαι) to the judgments of the mind. (*Antidosis* 180)

Compare Aristotle in the *Protrepticus*:

> Everything is well disposed when it is in accordance with its proper virtue, for to obtain this is good. Moreover, it's when a thing's most authoritative and most estimable parts have their virtue that it is well disposed; therefore the natural virtue of that which is better is naturally better. And that which is by nature more overarching and more deserving of command (ἀρχικώτερον καὶ μᾶλλον ἡγεμονικόν) is better, as a human is than the other animals; thus soul is better than body – for it is more overarching (ἀρχικώτερον γάρ) – as is the part of the soul which has reason and thought, for this kind of thing is what prescribes and proscribes and says how we ought or ought not to act. Whatever, then, is the virtue of this part is

III

In the first two chapters of the *Metaphysics* Aristotle contrasted and hierarchically arranged the theoretical and productive sciences, and in the parallel chapters of the *Nicomachean Ethics* he did the same for the practical and productive sciences. In the *Metaphysics* he groups the practical and productive sciences in order to strongly contrast these with the theoretical ones, offering a dichotomy. Elsewhere this same dichotomy is invoked, but the term "practical" is substituted for "productive," as in the opening of *Metaph.* 2.1: "Philosophy should be called knowledge of the truth. For the end of *theoretical* science is truth, while that of *practical* science is action; for even if they consider how things are, practical men do not study what is eternal but what stands in some relation at some time" (993b19–23, tr. Ross, adapted). For the purposes of this kind of dichotomous contrast the terms "practical" and "productive" can be used interchangeably, but the dichotomous contrast between these productive-practical and the theoretical sciences is stark. For one thing, theoretical sciences have eternal objects, but the practical-productive ones have only temporary things. Aristotle, surprisingly, suggests that the practical man can succeed without having any theoretical knowledge himself. It is vital to my argument, however, that this not be taken to imply that the practical man, such as the politician, can succeed in complete independence of theoretical philosophy – for it may well be that politicians depend on theoretical scientists (i.e. philosophers) as advisors, even if they do not themselves need theoretical knowledge. Nevertheless, when Aristotle argues against Socratic intellectualism he again draws a sharp contrast, this time using the term "productive" to stand in for both the practical and productive sciences in contrast to the theoretical:

> Socrates the elder thought that the goal was knowledge of virtue . . . Hence Socrates used to investigate what virtue is, but not how and by what means it comes about. His approach is applicable to the *theoretical sciences*, since there is nothing more to astronomy or to natural science or geometry than knowing or studying the objects of these sciences. Of course there is no reason that they should not also incidentally be useful to us for many of the necessities of life. But with the *productive sciences*, the goal is distinct from knowledge and understanding. For example, health is the goal of medicine,

> necessarily the most valuable virtue of all, both for everything in general and for us; in fact, I think one might actually take the position, that we are this part, either alone or especially. (*Protr.* 7.41.22–42.4, tr. Hutchinson and Johnson)

The continuation of this line of thought (in both Isocrates and Aristotle's *Protrepticus*) is a comparison between gymnastics and philosophy.

> and good order or something of that sort the goal of political science. Now it is indeed a fine thing to know each fine thing. Nonetheless, when it comes to virtue, knowing what it is is not the most valuable point, but understanding what brings it about. For we do not want to know what courage is, but to be courageous, nor to know what justice is, but to be just, as we want to be healthy rather than understand what being healthy is, and be in good physical condition rather than understand what being in good physical condition is. (*EE* 1.5, 1216b2–25, tr. Inwood and Woolf, adapted)

In such a context, where Aristotle operates with the traditional dichotomy between theoretical and "productive" science, there is no mention of the distinction between practical and productive sciences. When the important point is that a science aims at some result other than knowledge, whether a science is classified as productive or practical "makes no difference," as he puts it in *EN* 1.1, 1094a16. Theoretical sciences have no other end than knowledge of their objects, and any benefits they produce are incidental to the science. By contrast, productive sciences have as their end some result beyond knowledge, and thus these sciences must know not only what their objects are, but also how they are produced and used. Mere knowledge of the city, the citizens, political constitutions, etc. is not the end of political science, any more than mere knowledge of bodies and diseases is of medicine. In both cases the end is rather something else that is produced: health for the body, law and order for the body politic. Theoretical and practical uses of intelligence must therefore be sharply distinguished.[8]

Elsewhere Aristotle offers a trichotomous division of science: "this is classed as theoretical, practical, and productive, and each of these denotes a relation; for it theorizes about something, or produces something, or does something" (*Top.* 6.6, 145a15–18, tr. Pickard-Cambridge, adapted). This trichotomy is invoked in the beginning of *Metaphysics* VI and XI, where it is said to be an exhaustive division:

> In general every science which is intellective or at all involves intellect (πᾶσα ἐπιστήμη διανοητικὴ ἢ μετέχουσά τι διανοίας) deals with causes and principles, exact or indeterminate, but all these sciences mark off some particular being – some genus and inquire into this . . . Since natural science,

[8] Aristotle applies the distinction between theoretical and practical to operations of the mind in distinguishing between understanding and intelligence:

> Understanding is about the same objects as intelligence, but understanding and intelligence are not the same (οὐκ ἔστι δὲ τὸ αὐτὸ σύνεσις καὶ φρόνησις). For intelligence is directive, since its end is what should and should not be done; but understanding only judges (ἡ μὲν γὰρ φρόνησις ἐπιτακτική ἐστιν· τί γὰρ δεῖ πράττειν ἢ μή, τὸ τέλος αὐτῆς ἐστίν· ἡ δὲ σύνεσις κριτικὴ μόνον). (*EN* 6.10 = *EE* 5.10, 1143a6–10)

> like other sciences, confines itself to one kind of being (i.e. to that kind of substance which has the principle of its movement and rest present in itself), evidently it is neither *practical* nor *productive*. For the principle of production is in the producer – it is either reason or art or some capacity, while the principle of action is in the doer – viz. choice, for that which is done and that which is chosen are the same. Therefore if all thought is either *practical* or *productive* or *theoretical*, natural science must be *theoretical*. (*Metaph*. 6.1, 1025b5–9+18–26, tr. Ross, adapted; cf. *Metaph*. 11.7, 1063b36–4a19)

The trichotomy is especially useful when discussing the objects of the sciences, the particular genera that each science "marks off" and inquires into. In the most general terms, the productive sciences look to the principles employed by the *producers* in a given domain; the practical look to the principles employed by the *doers*. The theoretical sciences look to the principles employed by the *explainers* (or observers) of some particular being, such as a star, a number, a figure, or an animal. On the basis of this distinction between their objects, it is possible to rank-order not only the highest divisions of science (theoretical, practical, and productive), but also the highest divisions of theoretical science:

> There must be three theoretical philosophies, mathematics, natural science, and theology (since it is obvious that if the divine is present anywhere, it is present in things of this kind). And the highest science must deal with the highest genus, so that the theoretical sciences are superior to the other sciences (αἱ μὲν οὖν θεωρητικαὶ τῶν ἄλλων ἐπιστημῶν αἱρετώταται), and theology to the other theoretical sciences. (*Metaph*. 6.1, 1026a18–23, tr. Ross, adapted; cf. *Metaph*. 11.7, 1064b1–6)

As a result, we can formulate the following hierarchy:

(1) Theoretical Sciences:
  a. Theology (i.e. metaphysics, the science of first principles and causes)
  b. Mathematics
  c. Natural science
(2) Practical sciences
(3) Productive sciences

This entire network of evidence, then, indicates that Aristotle considers the theoretical sciences to be superordinate to the practical and productive sciences as a whole. The practical sciences are superordinate to the productive sciences. The theoretical sciences are superordinate to the practical, in turn, because (i) they are more general, and the causes and principles of the theoretical sciences govern the causes and principles of the practical sciences (including the cause for the sake of which), and (ii) because they are more final, since they are themselves intrinsically valuable, for the sake of the knowledge alone, and thus the productive and practical sciences are ultimately carried out for their sake.

IV

The picture that I have just drawn of a hierarchically arranged and highly interdependent set of sciences might be thought to be at odds with Aristotle's well-known critique of scientific unity. In the present section I will argue that, on the contrary, Aristotle's position is flexible enough to accommodate my interpretation, and I will attempt to show this by exhibiting the syllogistic structure to which all the proposed architectonic sciences should conform.

In the *Metaphysics* and *Topics* passages just discussed, Aristotle's insistence that each science covers only the objects in its own domain corresponds to a methodological prohibition on kind-crossing explanations:

> The items from which the demonstrations proceed may be the same; but where the kinds are different, as with arithmetic and geometry, you cannot attach arithmetical demonstrations to what is incidental to magnitudes – unless magnitudes are numbers. But I shall explain later how in some cases this is possible. (*APo.* 1.7, 75b2–6, tr. Barnes)

Let us consider the prohibition by contrasting two kinds of syllogistic explanation:

Syllogism 1 (valid)

1. A <a> B (geometrical principle)
2. B <a> C
3. A <a> C (geometrical fact)

Syllogism 2 (invalid; kind-crossing)

1. A <a> B (arithmetical principle)
2. B <a> C
3. A <a> C (geometrical fact)

Notice that in the above passage, Aristotle announces the prohibition but immediately points out the existence of exceptions to it (even possibly for the very example – arithmetic and geometry – that he gives to make his point). Later he describes in some detail at least four such exceptions.[9]

> The reason why differs from the fact in another way insofar as each is studied by a different science. These are the cases which are related to each other in such a way that the one falls under the other (θάτερον ὑπὸ θάτερον), e.g. optics to geometry, mechanics to solid geometry, harmonics to arithmetic, star-gazing to astronomy. Some of these sciences bear almost the same name as one another – e.g. mathematical and nautical astronomy, and mathematical and acoustical harmonics. Here it is for the empirical scientists to know the fact (τὸ μὲν ὅτι) and for the mathematical scientists to know the reason why (τὸ δὲ διότι). The latter possess demonstrations which give the explanations, and often they do not know the fact – just as people who study universals often do not know the particulars through lack of observation. The items in question are things which, being something different in their essence, make use of forms. For mathematics is concerned with forms; its objects are not said of any underlying subject – for even if geometrical objects are said of some underlying subject, still it is not as being said of an underlying subject that they are studied. Related to optics as optics is related to geometry, there is another science – namely the study of the rainbow. Here it is for the natural scientists to know the fact and for the students of optics – either of optics simpliciter or of mathematical optics – to know the reason why. (*APo.* 1.13, 78b34–79a13, tr. Barnes)

Although Aristotle usually refers to a simplified bipartite conception of subordinate sciences in his examples, this passage shows that he also developed a more complex tripartite conception of some of them.[10] In the bipartite version, a subordinate science describes the empirical fact (τὸ ὅτι), which the superordinate science is in charge of explaining by providing the reason why (τὸ διότι). In the tripartite version, an intermediate or mixed "mathematical-physical" science (e.g. optics) supplies a middle term that allows the application of the axioms of an autonomous mathematical science (e.g. geometry) to the explanation of a phenomenon described by empirical science (e.g. meteorology). Thus the first, independent science stands in a relation of architectonic superiority to two subordinate sciences in the following examples:

[9] Henry Mendell pointed out to me that Aristotle, although announcing his prohibition on kind-crossing in apparently strict theoretical terms, is in practice extremely casual in his own allowance of kind-crossing explanations.

[10] McKirahan 1978, 206–16 has persuasively argued that point. Both he and Hankinson 2005, 47–50 have pointed out difficulties with subalternate triples that I cannot here enter into. See also Lennox 1986, 42–44.

	Example 1:	Example 2:	Example 3:
Mathematical science:	1. Arithmetic	1. Stereometry	1. Geometry
Physical science:	2. Harmonics	2. Mechanics	2. Optics
Empirical science:	3. Acoustics	3. Levers	3. Meteorology

Let us examine in a little more detail the third example: an empirical science ("meteorology") supplies the facts, for example about lunar halos, that another science ("optics") explains by applying geometrical axioms. Both meteorology and optics are subordinate to a mathematical science ("geometry"), a science that possesses its own principles. Because these three sciences are properly arranged in this order of subordination, it is possible to construct a syllogism to explain a meteorological phenomenon, namely why the lunar halo is always a perfect circle:

Geometry:	1.	Circle <a> Figure with limits all equidistant to a single point
Optics:	2.	Figure with limits all equidistant <a> Light reflected by tiny, uniform mirrors
	3.	Circle <a> Light reflected by tiny, uniform mirrors (by 1+2)
Meteorology:	4.	Light reflected by tiny, uniform mirrors <a> Shape reflected in the clouds
	5.	Circle <a> Shape reflected in the clouds (meteorological fact, by 3+4)

The overarching major premise is simply an undemonstrated geometrical axiom (definition), the minor premises are theorems of optics or meteorology, and the conclusion is a description of the empirical phenomenon to be explained. Each proposition can be understood to belong to a different science, namely, geometry, optics, and meteorology, so long as these are understood to stand in these exact relations of subordination. The middle terms provide the causes (formal and efficient) of the phenomenon: it is because the light is reflected by tiny, uniform mirrors that the resulting figure is one with all its limits equidistant, and this in turn is the reason that the shape (the halo) always forms a complete circle.

This theoretical model is successful on its own terms, and was successfully applied by Aristotle to meteorology and by Aristotle (or his immediate successors) to the domain of mechanical problems, optics, and musical theory.[11] But there is of course plenty of room for doubt that anything like this can be made to apply outside these very close-knit mathematical-physical sciences to other theoretical sciences (such as metaphysics), or to these in relation to the productive-practical sciences. Certainly there have been interpretations of Aristotle's metaphysics, biology, and teleology that have not observed Aristotle's methodological restrictions on kind-crossing closely enough.[12]

V

In this section, I will nevertheless argue that in Aristotle's actual scientific practice, in the *Politics* at least, the subordination of politics to theoretical science is presupposed. In accordance with the structure just described, then, it must then be possible to demonstrate or explain political facts by means of natural scientific or otherwise theoretical principles. The challenge of the present section, then, is to show how one could possibly fill in the blanks of the following kind of syllogism:

[11] On optics and music, see McKirahan 1978. On meteorology, see Johnson 2009, and forthcoming, "Aristotelian Mechanistic Explanation." For an enlightening discussion of the influence of this model, see Lennox 1986.

[12] See the criticisms of Nussbaum 1978 in Kung 1982 for a well-argued example. Robert Bolton has recently argued that for Aristotle metaphysics and biology are wholly autonomous sciences, and that the results of neither science can place any constraints on the results of the other, and this fact has far-reaching implications for a number of recent studies that have unmethodically linked Aristotle's discussion of substances in both cases (Bolton 2010; see my review of the volume, Johnson 2011). A related case is the attempt to interpret Aristotle's teleology as anthropocentric on the basis of the *aporia* raised in *Phys.* 2.8 about whether regular winter rainfall is due to strictly necessary causes, or is for the sake of crop growth and human nutrition, which I have discussed at greater length in Johnson 2005. Were we to admit the anthropocentric teleological explanation as a theoretical proposition, we would have to find a way to subordinate meteorology (the science that states the facts about rainfall) to either the productive art of agriculture (rustic anthropocentrism) or to metaphysics (urbane anthropocentrism). But rainfall is explained by Aristotle on the basis of meteorological principles alone, which refer to the material and moving causes. There seems to be nothing insufficient about Aristotle's explanation of rainfall, and nothing to be gained by subordinating the moving-material explanation to a metaphysical principle. Thus there seems to be no justification here for violating the prohibition on kind-crossing. A more difficult case that I have also discussed in Johnson 2005 is *Pol.* 1.8, when Aristotle seems to infer from a principle of the biological sciences that "sustenance is provided by nature from birth to all," to the conclusion that "after the birth of animals, plants exist for their sake, and the other animals for the sake of humans" (1256b20–26). This inference cannot possibly be accepted as a principle of the biological sciences, since it is made in the context of a discussion of the acquisitive arts, which Aristotle explicitly says are subordinate to economics and hence politics (see 1256a5–10, where Aristotle distinguishes between two ways in which it may be understood to be subordinate). Since, as I will argue below, politics is itself subordinate to theoretical science (including natural, biological science), the acquisitive arts can only receive from, but not supply principles to, biological science, barring an inversion of the architectonic relation between these sciences (and thus a violation of the prohibition on kind-crossing). This is why there are no texts that support the anthropocentric notions in the biological works, and we only find such claims in the *Politics*.

1. A <a> B (theoretical principle of metaphysics, mathematics, or natural science)
2. B <a> C
3. A <a> C (political fact)

Indications in this direction come from a controversial passage from the *Protrepticus* that describes the relationship between politics, legislation, and natural science:

> Just as in the other productive skills the best of their tools were discovered on the basis of nature (for example, in the builder's skill, the carpenter's line, the standard ruler, the string compass) < . . . *a line of text is missing* . . . > for some are grasped with water, or with light and rays of sunshine, and it is by reference to these that we put to the test what is sufficiently straight and smooth to our senses – in the same way, the statesman must have certain criteria taken from nature itself, i.e. from the truth, by reference to which to judge what is just and what is good and what is advantageous . . . And in the other skills people pretty much know that they do not get their tools and their most precise calculations from the primary things themselves; they get them from what is second or third hand or at a distant remove, and they get their calculations from experience, whereas the imitation is of the precise things themselves only for the philosopher, for the philosopher's vision is of these things themselves, not of imitations. So just as no one is a good builder who does not use a standard ruler or any other such tool, but approximates them to other buildings, so too presumably if someone either puts laws to use in states or performs actions by looking at and imitating other human actions or political systems, whether the Spartan or that of the Cretans or of any other such state, he is neither a good lawmaker nor is he an excellent statesman; for an imitation of what is not beautiful cannot be beautiful, nor can an imitation of what is not divine and stable in nature be immortal and stable. But it is clear that the philosopher is the only producer to have both laws that are stable and actions that are correct and beautiful. For he is the only one who lives looking toward nature and toward the divine and, just as if he were some good navigator who hitches the first principles of his life onto things that are eternal and steadfast, he moors his ship and lives life on his own terms. (*Protr.* 10.54.22–56.2, tr. Hutchinson and Johnson)

In this passage, the model of architecture is used to construct an image of the architectonic superiority of philosophy over politics. The philosopher, who has theoretical knowledge, is said to be the only one of the "manufacturers" (δημιουργῶν) to produce secure laws and correct and beautiful actions. The philosopher is a good legislator (ἀγαθὸς νομοθέτης), and it is in fact his possession of *theoretical* knowledge that puts him in the position of an architect to command and control the subordinate arts, beginning with politics.

Werner Jaeger predicated a large part of his theory of Aristotle's development on his interpretation of this passage as "pure Plato" that "necessitates the dualist metaphysics of the forms as the theoretical basis."[13] Now to interpret the passage along Platonic lines is reasonable not only because the *Protrepticus* dates to Aristotle's Academic period, but also, more importantly, because of the Platonic style, as in the explanation: "for an imitation of what is not beautiful cannot be beautiful, nor can an imitation of what is not divine and secure in nature be immortal and secure." Here I set aside all questions of Aristotle's development and focus on the interpretation of the *Protrepticus* passage and its application to Aristotle's scientific practices in the *Politics*. For Jaeger the key to the comparison between the arts (i.e. architecture) and philosophy in the passage "comes from the account of the theory of Forms in the tenth book of the *Republic*."[14] Just as the carpenter of *Republic* x imitates not another couch or table, but looks to the ideal form of the couch or table, so the good architect imitates not another building, but looks to the ideal form of the building. Therefore the philosophical legislator should not imitate the form of existing states (like Sparta or Crete), but should look to the ideal form of the state. Such an interpretation of the *Protrepticus* passage could be supported by considering Aristotle's remark in *Politics* 4.1 that both legislation and politics must observe not only the "best" constitution, but also the one realizable in a particular political situation:

> In the case of a constitution also it belongs to one and the same science to examine all of the following. (1) Which is the best constitution? That is, what kind of constitution would be most desirable if there were no external hindrances to its realization? (2) Which constitution suits which persons? Since for many people it is perhaps impossible to achieve the best, the good legislator and the true politician must know both what is best absolutely and what is best in the circumstances. (*Pol.* 4.1, 1288b22–27 [in part], tr. Robinson, adapted)

One might ask, however, what would be the point of requiring two independent sciences to have parallel inquiries into both the ideal constitution and the most suitable constitution. The answer is that the two sciences are not independent; rather there is an architectonic structure according to which politics is subordinate to legislation. Aristotle makes this clear in

[13] Jaeger 1948, 90–91. [14] Jaeger 1948, 92.

the second common book of the *Ethics* in a way that sustains the analogy with architecture from the *Protrepticus*:

> Of the things concerned with the city, the intelligence which is architectonic is legislation (τῆς δὲ περὶ πόλιν ἣ μὲν ὡς ἀρχιτεκτονικὴ φρόνησις νομοθετική), while that which is related to this as particulars to their universal is known by the general name politics (πολιτική); this has to do with action and deliberation, for a decree is a thing to be carried out in the form of an individual act. This is why exponents of this art alone are said to take part in politics; for these alone act as manual laborers do (μόνοι γὰρ πράττουσιν οὗτοι ὥσπερ οἱ χειροτέχναι). (*EC* 2.8 = *EE* 5.8 = *EN* 6.8, 1141b24–29, tr. Ross, adapted)

Let us, then, consider how such an architectonic science might work in practice. Consider a sample syllogism built out of materials from *Pol.* 4.7, 1293b14–16:

Legislation (theoretical):	1. Aristocracy <a> Government with regard to wealth and excellence of whole populace
Politics (practical):	2. Government with regard to wealth and excellence of the whole populace <i> Carthage
Constitutional history:	3. Aristocracy <i> Carthage (political fact, by 1+2)

The third proposition belongs to an empirical science of politics that gathers facts about the history of political constitutions.[15] The second belongs to a political science that explains particular political facts by reference to an axiom (definition) of legislative science. Thus the reason Carthage (i.e. Carthage at a certain point in time) was an Aristocracy is that its constitution had concern for the wealth and excellence of the whole population. Although one would need to develop much more complex syllogisms than I have presented here in order to do justice to the scientific content of the *Politics* (e.g. with modal operators, careful attention to the quantities of the connectives, etc.), it is possible that such a logical structure could bring clarity to otherwise confused political matters. But whether or not it would do so, this experiment does show a way in which the *Protrepticus* passage can be interpreted along Platonic lines in a way consistent with Aristotle's later political science. The theoretical science

[15] The enterprise of a large-scale study of the history of political constitutions is mentioned at *EN* 10.9, 1181b12–22, and the *Ath. Pol.* is a surviving (fragmentary) example, whether it is by Aristotle or a follower. Substantial fragments of dozens more histories of constitutions that originate from the Peripatetic school survive; see Gigon 1987, 561–722.

grasps the form of the ideal state, and political science tries to bring about an imitation of this ideal state in a particular political situation – and not an imitation of an actually existing state, like Sparta. There thus appears to be an architectonic structure corresponding to a causal division of labor between (1) the legislator and architect, who supply the formal and final causes, and (2) the politicians and manual laborers, who supply the material and moving causes in order to bring about an imitation of these formal and final causes in a particular political situation.

The problem with such an interpretation, however, as Kurt von Fritz and Ernst Kapp pointed out in their introduction to Aristotle's *Constitution of Athens*, is that Aristotle does not actually say in the *Protrepticus* passage that the architect imitates the ideal form of the house, or that the legislator imitates the ideal form of political constitution, and in fact the technical language of the theory of Forms does not appear at all. Instead, Aristotle refers to the architect producing buildings by using accurate "instruments" calibrated not from other buildings but from nature itself. As von Fritz and Kapp say, the key questions are what nature means in this context, and what in political science corresponds to the accurate "instruments" of the architect.[16] In the rest of this chapter I will try to develop an answer to these questions.

But before doing so, I want to point out that, even though we cannot read the *Protrepticus* passage as "pure Plato" or as having anything to do with the theory of Forms, as Jaeger supposed, nevertheless the argument of the *Protrepticus* does conform to Platonic views in a more subtle way, considering the distinction of the *Politicus* between the "practical-manual" and the "directive-intellectual" sciences, which are also distinguished from the "calculative-intellectual" sciences. In what follows, I will argue that Aristotle seems to adopt the kind of hierarchy we wondered about in connection with the *Politicus*, according to which the architect directs the manual laborers, but is in turn subordinate to a higher non-directive ("critical") art, similar to calculation in that it is purely theoretical.

The good architect does not take the standards of straightness and smoothness by looking at other existing buildings, but by looking to nature, and calibrating instruments by means of the sun, water, etc. So too the good legislator does not take his standards of goodness and rightness by looking at other human actions and existing constitutions (like those of Sparta or Crete), but by looking to nature, and developing the standards by looking to natural (and "divine") things. As a result of knowing about these things, the good legislator is able to create "laws that are secure and

[16] Von Fritz and Kapp 1950, 34–35.

actions that are correct and beautiful." What is needed to make sense of this is to ask what kind of instruments legislators could possibly obtain from philosophy and natural science. And the answer stares us in the face in *Pol.* 1, which opens by criticizing "those who hold that statesman, king, household manager, and slave-master are the same" (1252a7–9), a view upheld by the interlocutors in Plato's *Politicus*, "supposing there were no difference between a large household and a small state" (1252a12–13; cf. Pl., *Polit.* 259b9–10). This position, Aristotle asserts "is not well put" (1252a9), and he promises to make the point clearer – the first stated purpose of the *Politics* – following a certain method. He then describes the method. Below I translate an extended passage – running over into the second chapter:

> The point will be clearer if we examine the matter according to the method that has guided us (κατὰ τὴν ὑφηγημένην μέθοδον). For *in the other cases* (ἐν τοῖς ἄλλοις) we have to analyze a composite into its irreducible elements (τὸ σύνθετον μέχρι τῶν ἀσυνθέτων ἀνάγκη διαιρεῖν), the smallest parts of the whole (ταῦτα γὰρ ἐλάχιστα μόρια τοῦ παντός). So let us in the same way examine the component parts of the state also, and we shall see better both how these too differ from each other and whether we can acquire some skilled understanding of each of the roles mentioned [sc. statesman, king, household manager, and master of slaves]. Now in this as *in other cases* (ὥσπερ ἐν τοῖς ἄλλοις) one would get the best view of things if one were to look at their natural growth from principles (ἐξ ἀρχῆς τὰ πράγματα φυόμενα βλέψειεν). *First: those which are incapable of existing without each other must unite as a pair.* For example, (a) male and female, for the sake of breeding . . . (b) that which naturally rules (ἄρχον δὲ φύσει) and that which is ruled (ἀρχόμενον) for preservation (διὰ τὴν σωτηρίαν). For that which can use intellect to look ahead (δυνάμενον τῇ διανοίᾳ προορᾶν) is by nature ruler and by nature master, while that which has the bodily strength to labor is ruled, and is by nature a slave. Hence master and slave benefit from the same thing. (*Pol.* 1.1–2, 1252a17–34, tr. adapted from Saunders)

Aristotle states that he will employ a method that is already familiar to the audience and, he seems to assume, presents no particular difficulties. And commentators have by and large treated Aristotle's remark as transparent and unproblematic, and so they have compiled references to other purported employments of this method in this and other books of the *Politics*, other works of practical science (such as the *Ethics*), or productive science (such as the *Poetics*), and of natural science (such as *On the Soul* and the biological works).[17] It seems to have caused little concern

[17] One who follows up all these references will find a great variety of "methods" being applied to a great variety of subject matters. A more systematic analysis of the expression κατὰ τὴν ὑφηγημένην μέθοδον throughout the corpus yields the same result: Aristotle uses this expression to refer to a variety of different local methodologies, some involving division, some part–whole analyses, and others entirely different methods.

that this apparently familiar Aristotelian methodology bears little prima facie resemblance to the prescribed method of the *Analytics*: there is no mention of syllogisms, middle terms, principles, causes, etc.; and in fact one might reasonably ask, as in so many other cases, whether the method employed in *Politics* I is not in tension with his views in the *Analytics*, where he expresses little enthusiasm for demonstration by division (*APr.* 1.31, *APo.* 2.5).

But of course Aristotle does recognize an important methodological role for division, as he states in *APo.* 2.13.[18] What he describes there seems to be a fair description of what Aristotle actually does in *Pol.* 1.1, if we do not cut off his description of the method at the end of the conventional chapter one, but take the beginning of chapter two to be a continuation of the description of the method, as seems reasonable from the fact that in both places Aristotle uses the expression ("in the other cases": 1252a18 and 25). Thus Aristotle analyzes the state into its atomic or indivisible elements: men, women, slaves. These items are then recombined into the pairs men/women and masters/slaves, according to the *metaphysical principle* that "those which are incapable of existing without each other must unite as a pair" – an indemonstrable axiom.[19] Aristotle then goes on to discuss the combination of these pairs into the triplet man/woman/slave, which generates the household (1259b9–15, oxen in lieu of slaves in poor households). Later he discusses the combination of households that generates the village (1252b15–27), and finally the combination of villages that generates the city (1252b27–1253a1). We can express these ideas into propositional and syllogistic form:

Metaphysics: 1. Combination <a> Pair that cannot exist in isolation

Natural science: 2. Pair that cannot exist in isolation <i> Male and female

[18] When you are dealing with some whole, you should divide the kind into what is atomic in form, i.e. into the primitives (διελεῖν τὸ γένος εἰς τὰ ἄτομα τῷ εἴδει τὰ πρῶτα), e.g. number into triplet and pair. Then you should try to get definitions of these items, e.g. of straight line and circle and right angle. After this, having got what the kind is, e.g. whether it is a quantity or quality, you should study its proper attributes through the primitive common items. For the characteristics of the items compounded from the atoms will be plain from the definitions, because definitions and what is simple are principles of everything, and it is of the simples alone that the characteristics hold in themselves – they hold of the other items in virtue of the simples. Divisions made according to differences are useful in this pursuit (αἱ δὲ διαιρέσεις αἱ κατὰ τὰς διαφορὰς χρήσιμαί εἰσιν εἰς τὸ οὕτω μετιέναι). (*APo.* 2.13, 96b15–26, tr. Barnes)

[19] Robert Bolton has pointed out to me the dubiousness of this "principle," a fair point that I do not here have the space to discuss.

Natural history:	3.	Combination <i> male and female (= family, from 1+2)
Natural science:	4.	Pair that cannot exist in isolation <a> Master and slave
Natural history:	5.	Combination <i> Master and slave (= natural slavery, from 1+4)
Economics:	6.	Pair that cannot exist in isolation <a> The male/female combination and master/slave combination
Economic history:	7.	Combination <i> Male/female combination and master/slave combination (= household, from 1+6)
Economics:	8.	Pair that cannot exist in isolation <a> Households
Economic history:	9.	Combination <i> Households (= village, from 1+8)
Economics:	10.	Pair that cannot exist in isolation <a> Villages
Political history:	11.	Combination <i> Villages (= city, from 1+10)

Political and economic facts, observed in political and economic histories or researches, are here explained, ultimately, by means of metaphysical and natural scientific principles. The overarching principle – itself indemonstrable – is that pairs that cannot exist in isolation must combine (or perhaps *should* combine for the sake of the good life – this raises complexities about the modal aspects of these syllogisms that I cannot enter into here). Other supposedly natural scientific principles, such as the axiom of "natural slavery" (#4 in the above scheme), are invoked to explain the fact of the emergence of familial and civilized life. These, I propose, are the kind of "instruments" taken from nature that the good legislator, like the good architect, must use in order to understand political reality and thus to construct laws that are secure and actions that are just.

Admittedly, this conception of politics is bound to appear excessively demonstrative and overly schematic as I have presented it here. I have tried to emphasize that I have only offered these as experimental examples, and I have to admit that they do not as yet get us very far into Aristotle's political science. But since Aristotle says in the second common book of the *Ethics* that science is a disposition or capability to demonstrate,[20] we should at least attempt to relate Aristotle's political science to the method

[20] *EN* 6.3 = *EE* 5.3, 1139b31–32.

of his *Posterior Analytics* if at all possible. Otherwise we will have either to revise our concept of science, or to stop thinking of political science as a science altogether. Neither of those are particularly attractive options. In fact, what is ultimately needed are *practical syllogisms* that either prescribe or justify certain kinds of action, that is, that explain certain actions, laws, or constitutional structures on the basis of ethical, economic, or political theorems. These theorems, however, will in turn have to be demonstrated on the basis of natural scientific and theoretical principles. Whether or not such a syllogistic conception of Aristotle's political science pans out, I have tried to show that an architectonic conception of philosophy in which theoretical science is authoritative is essential to any interpretation of Aristotle's practical sciences.

CHAPTER 10

Varieties of definition

David Sedley

[93b29]ὁρισμὸς δ᾽ ἐπειδὴ λέγεται εἶναι λόγος τοῦ τί ἐστι, φα/[30]νερὸν ὅτι ὁ μέν τις ἔσται λόγος τοῦ τί σημαίνει τὸ ὄνομα ἢ λό/γος ἕτερος ὀνοματώδης, οἷον τί σημαίνει [τί ἐστι] τρί/γωνον. ὅπερ ἔχοντες ὅτι ἔστι, ζητοῦμεν διὰ τί ἔστιν· χαλε/πὸν δ᾽ οὕτως ἐστὶ λαβεῖν ἃ μὴ ἴσμεν ὅτι ἔστιν. ἡ δ᾽ αἰτία / εἴρηται πρότερον τῆς χαλεπότητος, ὅτι οὐδ᾽ εἰ ἔστιν ἢ μὴ /[35] ἴσμεν, ἀλλ᾽ ἢ κατὰ συμβεβηκός. (λόγος δ᾽ εἷς ἐστὶ διχῶς,/ ὁ μὲν συνδέσμῳ, ὥσπερ ἡ Ἰλιάς, ὁ δὲ τῷ ἓν καθ᾽ ἑνὸς δη/λοῦν μὴ κατὰ συμβεβηκός.)

(**1**) Since a definition is said to be an account of what-it-is, it is clear that one sort will be an account of what is signified by the name or by some other name-like account, for example 'What does triangle signify?' When we grasp that this is we ask why-it-is. It is difficult to grasp in this way things of which we do not know that they are; and the reason for the difficulty has been stated previously,[1] namely that we do not even know if they are or are not, other than accidentally. (An account is 'one' in two senses: (a) by conjunction, like the *Iliad*, (b) by indicating one thing said of one thing, non-accidentally.)

[38]εἷς μὲν δὴ ὅρος ἐστὶν ὅρου ὁ εἰρημένος, ἄλλος δ᾽ ἐστὶν / ὅρος λόγος ὁ δηλῶν διὰ τί ἔστιν. ὥστε ὁ μὲν πρότερος σημαί/[94a1]νει μέν, δείκνυσι δ᾽ οὔ, ὁ δ᾽ ὕστερος φανερὸν ὅτι ἔσται οἷον / ἀπόδειξις τοῦ τί ἐστι, τῇ θέσει διαφέρων τῆς ἀποδείξεως./ διαφέρει γὰρ εἰπεῖν

One definition of definition is the above, but (**2**) another kind of definition is the account that indicates why-it-is. Hence the former signifies, but does not explain, whereas the latter will clearly be like a demonstration of what-it-is, although differing in its formulation from the demonstration. For there is a

My thanks to David Ebrey for critical comments on an earlier version.

[1] The reference is to 2.8, 93a24–29.

διὰ τί βροντᾷ καὶ τί ἐστι βροντή· ἐρεῖ / γὰρ οὕτω μὲν "διότι ἀποσβέννυται τὸ πῦρ ἐν τοῖς νέφεσι"·/ τί δ᾽ ἐστὶ βροντή; ψόφος ἀποσβεννυμένου πυρὸς ἐν νέφεσιν./ ὥστε ὁ αὐτὸς λόγος ἄλλον τρόπον λέγεται, καὶ ὡδὶ μὲν ἀπό/δειξις συνεχής, ὡδὶ δὲ ὁρισμός. (ἔτι ἐστὶν ὅρος βροντῆς ψό/φος ἐν νέφεσι· τοῦτο δ᾽ ἐστὶ τῆς τοῦ τί ἐστιν ἀποδείξεως συμ/πέρασμα.) ὁ δὲ τῶν ἀμέσων ὁρισμὸς θέσις ἐστὶ τοῦ τί ἐστιν / ἀναπόδεικτος.

difference between saying why it thunders and what thunder is. For one will state the former as follows: 'Because the fire is being quenched in the clouds'. And in reply to 'What is thunder?': a noise when fire is being quenched in the clouds. Thus the same account is being stated in a different way, and in the one case it is a continuous demonstration, in the other a definition.

(**3**) Also, a definition of thunder is 'noise in clouds': this is the conclusion of the demonstration of what-it-is.

(**4**) But the definition of unmediated terms is an indemonstrable positing of what-it-is.

ἔστιν ἄρα ὁρισμὸς εἷς μὲν λόγος τοῦ τί ἐστιν ἀναπό/δεικτος, εἷς δὲ συλλογισμὸς τοῦ τί ἐστι, πτώσει διαφέρων / τῆς ἀποδείξεως, τρίτος δὲ τῆς τοῦ τί ἐστιν ἀποδείξεως συμ/πέρασμα. φανερὸν οὖν ἐκ τῶν εἰρημένων καὶ πῶς ἔστι τοῦ τί / ἐστιν ἀπόδειξις καὶ πῶς οὐκ ἔστι, καὶ τίνων ἔστι καὶ τίνων οὐκ / ἔστιν, ἔτι δ᾽ ὁρισμὸς ποσαχῶς τε λέγεται καὶ πῶς τὸ τί / ἐστι δείκνυσι καὶ πῶς οὔ, καὶ τίνων ἔστι καὶ τίνων οὔ, ἔτι δὲ / πρὸς ἀπόδειξιν πῶς ἔχει, καὶ πῶς ἐνδέχεται τοῦ αὐτοῦ εἶναι / καὶ πῶς οὐκ ἐνδέχεται.

Hence definition is (**4**) an indemonstrable account of what-it-is, (**2**) a deduction of what-it-is, differing in linguistic form from the demonstration, (**3**) thirdly the conclusion of the demonstration of what-it-is.

From what has been said it is clear both in what sense there is demonstration of what-it-is and in what sense there is not, and of what things there is and of what things there is not; also in how many senses definition is spoken of, and in which of those it explains what-it-is and in which it does not, and of what things there is definition and of what things not; and also how it is related to demonstration, and in what ways it admits of being of the same thing, and in what ways not.

With this notoriously condensed chapter on definition, Aristotle concludes a major section of *Posterior Analytics* book 2. Hence in what my translation marks off as the final paragraph, he is summarizing not just chapter 10 itself but the whole of chapters 8–10. Nevertheless, a portion of that summary is devoted to chapter 10's own concluding remarks on definition, and among other things he claims to have shown how many kinds or senses of definition there are (94a16). How many?

Answering this question is not merely a matter of arriving at the correct number, for among the kinds of definition thought to be present in the chapter is one, often called 'nominal definition' (as opposed to 'real' or 'substantial' definitions), for which this passage is, in its first sentence, a virtually unique source, and which has exercised a considerable influence within and beyond the Aristotelian tradition. The chapter seems, that is, to be Aristotle's one explicit acknowledgement, among the types of definition he recognizes, of a primarily lexicographical approach to word meaning, whereby a word's definition is supplied simply by a phrase synonymous with it, without reference to the way things actually stand in the world.[2] Thus, it is said, if there is a definition of 'goat-stag', it can only be a nominal definition, there being absolutely no such thing as what a goat-stag is (cf. *APo.* 2.7, 92b5–8).

When we set out to count the number of definitions of definition listed in chapter 10, we find that on the face of it there are four. I have added bold numbers to the above translation in order to identify these. (The translation follows the OCT text placed alongside it, except that I have sometime opted in the translation for a different punctuation and paragraphing.) In the initial listing (93b29–94a10) the four kinds are:

(1) a nominal definition (as I shall call it for now, following convention);
(2) a causal definition, equivalent in informational content to an entire demonstration, e.g. thunder as 'a noise when fire is being quenched in the clouds';
(3) a non-causal definition, equivalent in informational content to the conclusion of a demonstration, e.g. thunder simply as 'noise in clouds';
(4) an indemonstrable definition of some first principle of a science.

[2] For the purposes of this chapter I am adopting the traditional understanding of what 'nominal' definition means. According to some interpreters (e.g. Bolton, and Demoss and Devereux, as cited below) nominal definitions do in fact have existential import, even if they do not fully capture the essence of the *nominatum*. In my view these interpreters are right to insist on existential import in the main cases they discuss, but misleading in retaining the term 'nominal' for them.

Yet when Aristotle recapitulates the list (94a11–14), there are only three, and these are quite clearly, in their new order (presumably a descending order of precision), recognizable as (**4**), (**2**) and (**3**). What then happened to (**1**)? Solutions to this recognized problem divide broadly into three types.

It has been suggested by Jonathan Barnes[3] that Aristotle's summative list of three does not claim to be exhaustive, so that there is no conflict with the initial list of four, which should therefore be allowed to stand. Although this is technically correct, it is hard to accept that a merely selective recapitulation could take the form 'Hence definition is x, y and thirdly z' (94a11–14), which gives every impression of being a complete enumeration – an impression confirmed when he goes on immediately to conclude the chapter by telling us that he has now said how many kinds of definition there are (94a16). Moreover, these three types of definition were already listed, in the same order **4–2–3,** and with an equally strong appearance of exhaustiveness, back at *APo.* 1.8 (75b31–2): '. . . since definition is either a starting point of a demonstration [= (**4**)], or a demonstration differing in its formulation (θέσει) [= (**2**)], or a kind of conclusion of a demonstration [= (**3**)]'.

A second option, widely favoured and traceable back at least to Themistius,[4] is that the first item, nominal definition, is excluded from the final tally because it is not really a proper definition at all. This too looks strained, when we bear in mind Aristotle's concluding reference to this first type of definition at 93b38: 'One definition of definition is the above, but another kind . . . '.[5]

A third solution, due to Zabarella, Ross and Pellegrin among others, is to propose that the third type of definition and the first are in fact one and the same.[6] In what follows this will emerge as much the strongest contender,[7] but it will need to be approached indirectly, by first comparing the types of definition that I have labelled (**2**)and (**3**).

[3] Barnes 1993, 222–3. [4] Themistius, *In An. Post.* 51.23–24.

[5] A more subtle variant, due to Charles 2000, esp. 26–33, is to discount nominal definition on a slightly different ground: it is a sort of definition, sure enough, but unlike the other three it does not define a kind. It is beyond the scope of my short chapter to respond to Charles' elaborately defended interpretation, beyond pressing the merits of a different reading.

[6] Zabarella 1594; Ross 1949, 634–5; Pellegrin 2005, 408–9. I particularly applaud Pellegrin's very brief but incisive comments on the chapter. Despite differences on many aspects, the same basic identification of (**1**) and (**3**) is at least implicitly endorsed by Bolton 1976, Demoss and Devereux 1988; cf. also Detel 1993, 2.670–71, 676.

[7] I acknowledge the awkwardness, emphasized to me by David Ebrey, of taking ἔτι (94a7), 'Also . . . ', to reintroduce a type of definition that has already been listed once. The reason, I take it, will be that it is only at this second occurrence that the type is formally characterized in comparison with the other two, namely in terms of three different deductive statuses (94a1–10).

Types (2)–(4) are distinguished in terms of the different inferential modes by which, at least notionally, they are or are not deducible. Type (4), being the definition of a primary concept in some domain, does not represent the outcome of any deductive process at all. How about the other two?

A type-(2) definition conveys the cause of the *definiendum*, as in the example 'Thunder is a noise when fire is being quenched in the clouds.' Hence this formula encapsulates the causal information supplied by an entire demonstrative deduction, perhaps expressed along the following lines:

> When fire is quenched, there is noise;
> Fire is quenched in clouds;
> Therefore there is noise in clouds

The definition, thus implied, of thunder as 'noise when fire is quenched in clouds' differs from any such demonstrative deduction in its linguistic form (its πτῶσις, 24a12; its θέσις, 94a2), but conveys the same causal explanation.

A type-(3) definition is, it seems, the conclusion of that very same demonstration, but this time taken on its own, without embodying the causal information that precedes. The example given is the simple 'Thunder is noise in clouds.' In Chapter 8, at 93a9–15, the same notion of definition was more cautiously specified as the conclusion, not necessarily of a demonstration, but of any 'logical deduction of what-it-is ' (λογικὸς συλλογισμὸς τοῦ τί ἐστιν). We will need to bear this looser characterization in mind in what follows.

We may hope to make some progress with illuminating the use of these type-(3) definitions by appeal to two passages in other works where the same diagnosis appears to be being applied to specific cases of definition – and especially the second of these passages, one whose significance has not as far as I know been appreciated.

At *de Anima* 2.2, Aristotle advertises the merits of developing a properly causal account of soul. To explain the difference between non-causal and causal definitions, he writes in Chapter 2:

> For the definitional account should not only indicate the that (τὸ ὅτι), as most definitions do, but the cause too should be contained in it and be made evident. At present the accounts of definitions are like conclusions (ὥσπερ συμπεράσματα). For example: 'What is squaring? It is for there to be an equilateral right-angled figure equal to an oblong one.' That kind of definition is an account of the conclusion. But one that says that squaring is the discovery of a mean proportional states the cause of the thing.

Here Aristotle presupposes and applies what is recognizably the *APo.* 2.10 notion that a non-causal definition corresponds to a mere 'conclusion' of a demonstration, but this time with the added clarification that such a description is no more than an analogy: a non-causal definition is 'like' a conclusion. The clarification absolves us of having always to identify the precise deduction whose conclusion it is, and the same latitude can perhaps be assumed in the *APo.* as well. We should note too that the specimen non-causal definition offered in the *DA* passage might easily have been thought of as merely 'nominal', supplying no more than the lexical meaning of 'squaring', without determining how if at all such a procedure might be realized. Yet it is meant by Aristotle to be comparable in status to 'most definitions', apparently including the very substantive, and by no means merely nominal, non-causal definition of soul defended in chapter 1 of the same book, 'the first actuality of a natural organic body'. This casts some doubt over whether Aristotle ever intended a contrast between type-(**3**) definitions, which state what the *definiendum* is but in non-causal terms, and merely 'nominal' definitions as traditionally found in type (**1**). I shall be arguing that he did not, and that the notion of 'nominal' definition may in fact have no established place at all in his theory of definition.

An equally relevant but less appreciated passage is the beginning of *Physics* 4.7, on void. Having sketched the arguments for and against the existence of void (chapter 6), Aristotle continues as follows (213b30–214a4):

> In relation to the question which of the two views is right, one has to grasp what the name signifies (τί σημαίνει τοὔνομα). Void is thought to be place in which there is nothing. The reason for this is that they think that what-is is body, that all body is in a place, and that place in which there is no body is void, so that if somewhere there is no body, there is nothing there. Again, they think all body is tangible, and that what is tangible is whatever has heaviness or lightness. Hence it emerges from a deduction (ἐκ συλλογισμοῦ) that void is that in which there is nothing heavy or light. This, as we said before as well, emerges from a deduction.

This time there is some emphasis on an actual deduction – not, we should note in passing, a 'demonstration' – presumed to be used by the void theorists in arriving at the putative definition of void. (The second reference to a deduction, in the closing sentence, does not despite appearances refer back to any preceding chapter of the *Physics*, but to the immediately preceding sentence.) What is the deduction? Apparently a two-stage one. 'Again' (πάλιν, 213b34) marks the transition to the second stage, in which the conclusion of the first stage serves as an implicit premise:

Stage 1
Void = place in which there is nothing
What is = body
Therefore void = place in which there is no body
Stage 2
Body = tangible
Tangible = heavy or light
Therefore void = that in which there is nothing heavy or light

Both stages would no doubt need some reworking to shape them into a formally correct pair of Aristotelian deductions – always supposing that Aristotle would want to, given that he does not endorse all the premises. But one more immediate significance of the passage is that, unlike the *APo.* and *DA* parallels, it confirms that a definition may indeed emerge as the actual conclusion of (and not merely as analogous to the conclusion of) of a piece of deductive reasoning.

As Aristotle proceeds to make explicit, the two stages result in two distinct definitions of 'void'. Void could not include geometrical points within its definition, he remarks (214a3–6); on the other hand, he continues (214a6–12),

> it does appear[8] that void is spoken of in one sense as that which is not full of body perceptible by touch, where what is 'perceptible by touch' is what has heaviness or lightness . . . and in another sense as that in which there is not some particular or corporeal substance.

These two definitions are unmistakably, in reverse order, the two that resulted from stages 1 and 2 above. The first (in the original order), 'Void is place in which there is no body', is the one that will be put to work in the extended series of critiques that now follows (first at 214a16–17). But what is the function of the further definition inferred from it, 'Void is that in which there is nothing heavy or light'? In the current polemical context this latter is the less obviously motivated of the two definitions, yet it has now been formally recorded as at least an ostensible definition of void. How are we to account for its inclusion?

We may approach that question by first asking another. Where do these two putative definitions of void belong in Aristotle's list of definition types? Both are non-causal, so cannot be of type (**2**). Nor, furthermore, is either a

[8] ἀλλ' οὖν (214a6), on which cf. Denniston 1950, 442–43: 'Following the rejection of a suggestion . . . Usually in such cases ἀλλ' οὖν introduces a more moderate suggestion . . . '.

definition of a first principle, so they cannot be of type (4). Are they, then, of type (1) or type (3)?

It is admittedly here if anywhere that it becomes plausible to see Aristotle as invoking merely nominal definitions, in accordance with the traditional interpretation of type-(1) definitions. In his own view, after all, there is no such thing as 'void'. So in introducing *Phys.* 4.7 with the formula 'one has to grasp *what the name signifies*' (τί σημαίνει τοὔνομα) (213b30–31), rather than 'what void is', he really does seem likely to have chosen a wording that avoids any commitment to there actually being void, *however* defined. And according to commentators like Themistius and Philoponus,[9] in fact, it is paradigmatically in order to be able to handle the definitions of unreal items like 'void' and 'goat-stag' that Aristotle needs a theory of merely nominal definition in the first place.

On the other hand, following this seemingly nominal formulation of his definitional question, we saw Aristotle proceed to present the two purported definitions of void as being of type (3), the conclusion of a (non-causal) piece of deductive reasoning. This casts serious doubt on the idea that type-(1) definitions were ever intended to be marked off, as merely nominal, from type-(3) definitions. And this replicates a doubt that also emerged from our examination of *DA* 2.2.

In the light of that doubt, we can now return better equipped to the question why in *Phys.* 4.7 Aristotle deduces two working definitions of void, the second of which he does not appear to need for the ensuing critique. A likely, but unappreciated, reason is that the second putative definition of void as 'that in which there is nothing heavy or light' is in fact *not*, for Aristotle, a nominal definition, that of a purely fictional entity like a goat-stag. For at the end of his critique of void, in *Phys.* 4.9, where he concludes that there is no void, he adds, 'unless someone wants to stick with calling the cause of motion "void": in this way the matter of the heavy and the light would, qua such, be the void' (217b21–24). In these words he may not seem to have given his own endorsement to the suggested legitimate usage of 'void'; but that impression is corrected a few lines later when he sums up his entire discussion of void: 'Concerning void, let the above suffice for distinguishing the senses in which it exists and in which it does not' (217b27–28). Thus, despite the overwhelmingly hostile reception that he has given to the notion of void in the intervening chapters, Aristotle is in the end at pains to concede that the term nevertheless does in a way signify something real.

[9] Themistius, *In An. Post.* 51.6–7; Philoponus, *In An. Post.* 359.25–360.9.

This signification is, moreover, recognizably dependent on the second putative definition that was deductively obtained at the beginning of chapter 7, but then set aside during the intervening critique. That definition was that void is 'that in which there is nothing heavy or light', and the acceptable sense of 'void' at the end of his discussion turned out to be 'the matter, qua such, of the heavy and the light'. This resemblance is hardly an accident. If void is held, explicitly or implicitly, by its proponents to be what contains neither the heavy nor the light, that belief can be recognized as gesturing towards a correct definition according to which void is the matter underlying the heavy and the light, and, qua their matter, distinct from them both.[10] The methodological point, however condensed, is the characteristically Aristotelian one of showing that even a seriously incorrect *endoxon* contains a germ of the relevant truth.

It may still be objected that at any rate the *first* definition of void, as place empty of all body, could not be more than nominal, since there is in reality nothing for it to correspond to. In reply, we must attach significance to the fact that the second definition was represented by Aristotle as derived deductively from this first one. That is, those who postulate void as a place empty of all body are in fact, *whether or not they fully realize it*, implicitly postulating it as 'that which is neither heavy nor light'. If the latter definition maps onto some feature of reality, so too, at least indirectly, does the former.

This understated but clearly detectable strategy illustrates, I suggest, why in his physical investigations Aristotle simply does not worry that the object of this or that inquiry – here void, and in the preceding book of the *Physics* the infinite – may prove to be as illusory as a goat-stag, and hence that its definition may be condemned to being merely nominal. The insight thus provided may also help us to understand the following remark from *APo.* 2.7 (92b26–30):

> If then one who defines explains either what-it-is or what the name signifies, if a definition is *not in any way* (μηδαμῶς) of what-it-is it would be an account signifying the same thing as the name. But that would be absurd. For, first of all, there would be definition both of non-substances and of non-existent things (for it is possible to signify non-existent things as well) . . .[11]

[10] This understanding of void as 'the matter, qua such, of the heavy and the light' is to be distinguished from its identification as 'the matter *of body*', which Aristotle rejected back at 4.7, 214a13–16.

[11] εἰ ἄρα ὁ ὁριζόμενος δείκνυσιν ἢ τί ἐστιν ἢ τί σημαίνει τοὔνομα, εἰ μὴ ἔστι μηδαμῶς τοῦ τί ἐστιν, εἴη ἂν ὁ ὁρισμὸς λόγος ὀνόματι τὸ αὐτὸ σημαίνων. ἀλλ' ἄτοπον. πρῶτον μὲν γὰρ καὶ μὴ οὐσιῶν ἂν εἴη καὶ τῶν μὴ ὄντων· σημαίνειν γὰρ ἔστι καὶ τὰ μὴ ὄντα.

Here the alternative of ending up with a *merely* nominal definition – a definition that is 'not in any way' of what-is – is dismissed as absurd. Why so? It may, we can now see, be because all definitions, when as fully developed and scrutinized as the definition of void was, are assumed to find some correspondence in objective reality. Although Aristotle regularly (including, in 2.10, 93b32–35) presents the 'Does it exist?' question as logically prior to the 'What is it?' question, he does not in practice have to worry about the danger of formulating definitions that may altogether fail to pick out existent things, because even the most contentious items in his own inquiries, well exemplified by void, are *not* altogether non-existent.

When, in the light of these findings, we return to *APo.* 2.10, we may legitimately doubt whether Aristotle meant to include in his list not only real definitions, which say of a thing what-it-is, but also merely nominal definitions, which restrict themselves to saying what the word signifies. On a little further inspection, it becomes clear that this chapter's type-(**1**) definitions are no more distinct from type-(**3**) definitions than they have proved to be in *DA* 2.2 and *Phys.* 4.7.

For confirmation, note first the opening two lines of the chapter: '*Since* a definition is said to be an account of what-it-is, it is clear that one sort will be an account of what is signified by the name or some other name-like account.' It is *because* a definition is, generically, an account of what-it-is that we should expect one kind of definition to be an account of what is signified. True, he does not in these words make it entirely unambiguous that this account of what is signified is itself a definition, but he will do so when referring back to it at 93b38: 'One definition of definition is the above.' It seems then that saying what '*x*'[12] signifies does not stand in any sort of contrast to saying what *x* is, but is itself a way of saying just that.

Secondly, at 93b31–32 the proffered example of this kind of definition, 'What does triangle signify?', where the sample *definiendum* is something Aristotle immediately acknowledges does exist, would be poorly suited to illustrating a *merely* nominal definition.[13] And a textual doubt can be added here. The text as quoted above, οἷον τί σημαίνει [τί ἐστι] τρίγωνον,

[12] I add the quotation marks here and elsewhere with some hesitation, because Aristotle's own uses of σημαίνειν do not typically distinguish between use and mention: cf. *APo.* 1.1, 71a14–16, 1.10, 76a32–34, etc.

[13] Pellegrin 2005, 408, aptly comments: 'Nous ne sommes donc pas dans un cas comme celui du bouc-cerf, dont il n'y a pas de définition au sens propre puisqu'il n'existe pas. Ce premier type de définition du chapitre 10 – qu'on a la malheureuse habitude d'appeler "définition nominale" – est donc une vraie définition qui a un référent réel.'

incorporates a deletion proposed by Ross. Clearly from the grammatical point of view Ross is right that the two questions, τί σημαίνει and τί ἐστι, cannot simply be juxtaposed in the way the transmitted text presents them.[14] But if one of them has to go, why delete τί ἐστι, and not rather τί σημαίνει? The latter phrase is much the likelier of the two to have crept in as a gloss or attempted correction. For from antiquity onwards this passage was regarded as the *locus classicus* on nominal definition, for the simple reason that its expression λόγος . . . ὀνοματώδης (30–31) was long misconstrued as being Aristotle's own term for 'nominal definition'.[15] An ancient reader or scholar approaching the text with that expectation was likely to be perplexed to read λόγος ἕτερος ὀνοματώδης, οἷον τί ἐστι τρίγωνον, believing this to fly in the face of what a 'nominal definition' should be. The addition or substitution of τί σημαίνει may well be a consequence.

If we retain τί ἐστι and instead delete τί σημαίνει,[16] we get the following account of a type-(**1**) definition:

> Since a definition is said to be an account of what-it-is, it is clear that one sort will be an account of *what is signified by* the name or by some other name-like account, for example '*What is* a triangle?'

On this reading – supported by what we have witnessed in *DA* 2.2 and *Phys.* 4.7 – there is no intended distinction between a type-(**1**) and a type-(**3**) definition. To say what *x*'s name signifies *just is* to say what *x* is. Aristotle has chosen to use the former locution, I suspect, for no more interesting reason than that he wants to be precise here in demarcating what kind of term can serve as *definiendum*. A *definiendum* is typically expressed by a name/word/noun (ὄνομα), but can also, as he adds here, be expressed by what in later grammatical terminology could be called a noun phrase and here in Aristotle's own terminology is called 'some other name-like

14 Pellegrin retains the MSS reading οἷον τὸ τί σημαίνει τί ἐστι ᾗ τρίγωνον, translating 'par exemple ce que signifie <la réponse à la question> "qu'est-ce qu'un 'triangle' en tant que triangle?"' Similarly Detel, 1993, 1.69 translates 'was es bezeichnet, was ein Dreieck ist'. I am inclined to agree with Ross 1949, 636, that this reading 'seems impossible'.

15 Galen, *Diff. puls.* 4, 8.705.11–12; Themistius, *In An. Post.* 51.3–7; Philoponus, *In An. Post.* 372.8–19. 93b31–2 is then taken to mean 'one sort will be an account of what is signified by the name, i.e. a nominal account'. As recent commentators point out, this construal has to ignore ἕτερος in 31.

16 I owe this suggestion to Malcolm Schofield. Ross cites Philoponus, *In An. Post.* 372.17–18, οἷον παριστᾷ τί σημαίνει τὸ ὄνομα τοῦ καθὸ τρίγωνον, as evidence that his copy contained τί σημαίνει at this point, but it is at least as likely to be evidence of the (incorrect) interpretation that led to the intrusion of these words (Ross 1949, 636).

account'.[17] As Ross points out,[18] 'straight line' would be a good example of such a *definiendum*.

Whether expressed in terms of what *x is* or of what *x*'s name *signifies*, this kind of definition can be represented as, or compared to, a bare conclusion of deductive reasoning (type-(**3**)), by contrast with those more explanatory definitions that have the causal premise of a demonstrative deduction built into them (type (**2**)). Such is Aristotle's main contention in *Apo.* 2.10, and it has nothing to do with any theory of 'nominal definition'.

[17] The precision reflects his concern, in 2.7 (92b30–32) and again in the excerpted passage of 2.10 (93b35–37), to distinguish which kinds of *logos* can and which cannot be used in definition. His main concern is that the *definiens* should have the right kind of non-accidental unity, but there is no reason to doubt that in chapter 10 he is applying the same concern to the *definiendum* as well.

[18] Ross 1949, 635.

CHAPTER 11

Empty words

Sean Kelsey

There is no Body in the Commonwealth of Learning, who does not profess himself a lover of Truth: and there is not a rational Creature that would not take it amiss to be thought otherwise of. And yet for all this one may truly say, there are very few lovers of Truth for Truths sake, even amongst those, who perswade themselves that they are so.

– Locke, *Essay* IV.xix.1 (Nidditch)

Aristotle sometimes criticizes an 'argument' or 'theory' (λόγος) for being 'empty' (κενός), charging its proponents with 'empty talk' (κενολογία, κενολογεῖν, κενῶς or διὰ κενῆς λέγειν).[1] This will serve as a preliminary specification of my topic, which is a particular failing Aristotle sometimes uses this vocabulary to mark, namely, a certain lack of seriousness in philosophical or scientific work, which manifests itself in the propounding of theories and arguments without regard for their truth.

One might think that, in Aristotle's view, such a failing must be rather the exception, too infrequent to merit his sustained, critical attention, and thus that his views on the topic can hardly be of more than marginal interest or importance. On the contrary, as I hope to show, the phenomenon of 'empty

I would like to thank Gavin Lawrence and Henry Mendell for the opportunity to present a draft of this chapter at UCLA, and those present for their questions, comments, and criticisms. For written comments on an earlier draft I would like to thank David Ebrey, Chris Frey, Doug Hutchinson, Joe Karbowski, André Laks, Jonathan Lear, Yannig Luthra, Alasdair MacIntyre, Ian McCready-Flora, Stephen Menn, Allison Murphy, Dave O'Connor, and Michael Pakaluk. A special thanks to Joe Karbowski, Yannig Luthra, and Dave O'Connor, for many hours of helpful and stimulating discussion. In thinking about this topic I have drawn heavily on the writings of J. H. Newman (most notably his *Grammar of Assent*, but see too (e.g.) *Parochial and Plain Sermons* 1.10, 5.3, 5.16; *Discussions and Arguments on Various Subjects* IV.7, *Loss and Gain: The Story of a Convert*, *Fifteen Sermons Preached before the University of Oxford* 4); indeed, one idea of the project was to try to find in Aristotle the 'Aristotelian' I find in Newman. (The passage from Locke I use as an epigraph is quoted by Newman in *Generation of Animals* II.1.)

[1] See e.g. *GA* II 8, 747b27–748a16; *EE* I 8, 1217b16–23; *Metaph.* A 9, 991a19–23, 992a24–29; M 4, 1079b3–11; M 5, 1079b23–27; *Sens.* 2, 437b11–23; *Rh.* III 13, 1414b13–18; also *de An.* I 1, 402b22–403a2; *PA* I 4, 644b15–20.

talk' is positively a theme in Aristotle, one that attracts his attention not only because of how prevalent he thinks it is (which already is something), but also because of how commonly he thinks it passes unnoticed by both speaker and audience alike. It is this combination that makes Aristotle's views on this topic so interesting; in his view, a proper regard for the truth, though absolutely essential to good scientific practice, is not only very rare, but also seldom missed.

Though my ultimate interest here is in Aristotle's natural science, I will begin with passages from his ethical writings. I do so for two reasons. First, it is here that we find, personified in nameable types, the full panoply of possibilities implicit in the theme of this volume: the pseudo-scientific 'professor' (σοφιστής), peddler of theories but innocent of practice, the unscientific 'politician' (πολιτικός), seasoned in practice but downright suspicious of theory, and, finally, the 'legislator' (νομοθετικός), consummate statesman and true 'scientist', in whom theory and practice are properly combined. These ethico-political types are a useful point of reference in working out Aristotle's views about the pathologies that come of divorcing 'theory' from 'practice' in natural science. Second, as we will see, the specific pathology I am interested in is as much moral as it is intellectual; that is, it results from a lack of intellectual character, and manifests itself in what is not so much bad reasoning as pretend reasoning, inasmuch as it is not only propounded but also conducted and concluded without regard for the truth. This aspect of the phenomenon is comparatively easy to spot in ethical and political contexts; starting from such contexts will make it easier to see how often Aristotle complains of this failing, even in his scientific writings.

The topic is a large one, with a past and a future beyond Aristotle, and it is bound up with other large topics even within Aristotle: for example, with his conceptions of 'experience' (ἐμπειρία), 'education' (παιδεία), 'science' (ἐπιστήμη), and 'trust' (πίστις), with various relationships among these, and with how they interact with a contrast he sometimes draws between 'word' (λόγος) and 'object' (πρᾶγμα) or 'deed' (ἔργον). Accordingly I do not pretend to offer anything like a definitive treatment, but rather just to make a first foray into relatively unexplored territory; my intention is to start a conversation, not to finish one, and I would be glad if others saw fit to enlarge, refine, or correct any of the things I say here.

I

The first passage I want to consider is taken from *Eudemian Ethics* I 6. The chapter as a whole is given to justifying the character of the ensuing discussion, which despite the practical nature of its topics, will nonetheless be 'scientific' or 'philosophical'.[2] The passage is of interest here for its suggestion that purveyors of empty talk are merely pretending at philosophy or science.

Aristotle begins the chapter by explaining that his aim is not merely to persuade, but to seek 'conviction' or 'proof' (πίστις) specifically 'through reasons' (διὰ τῶν λόγων), that is, causes (αἰτίαι or τὸ διὰ τί) (1216b26–27, 35–39, 1217a10–15). However, he cautions, this is something we need to be very careful about; a show of reasons is not enough to make an argument scientific: reasons vary by discipline, and so too do arguments and theories that are and are not 'philosophical' (1216b35–40). He continues:

> Because it seems to be the mark of a philosopher to say nothing at random, but always with reason, people often pass unnoticed giving reasons (λέγοντες . . . λόγους) that are irrelevant to the subject at hand and empty (ἀλλοτρίους . . . τῆς πραγματείας καὶ κενούς). This they do sometimes through ignorance, sometimes through affectation (ὁτὲ μὲν δι' ἄγνοιαν, ὁτὲ δὲ δι' ἀλαζονείαν), with the result that even men of experience and practical ability are taken in by them, though they themselves neither possess nor are capable of executive or practical thought (διάνοιαν ἀρχιτεκτονικὴν ἢ πρακτικήν). This happens to them thanks to a lack of education (δι' ἀπαιδευσίαν); for lack of education in a subject is the inability to discriminate between reasons that are relevant to the subject from those that are irrelevant (τούς τ' οἰκείους λόγους τοῦ πράγματος καὶ τοὺς ἀλλοτρίους). (1216b40–1217a10)

In this passage Aristotle is primarily concerned with the danger of being taken in by empty talk; this is in keeping with his larger agenda, which is to justify his giving practical topics like virtue and happiness a scientific treatment (cf. 1216b16–19, 36–38). However, he also touches on the causes of empty talk (sometimes affectation, sometimes ignorance), and the collocation of these causes is suggestive. When the result of affectation, empty talk verges on being a deliberate counterfeit of philosophical discourse; the speaker pretends to treat his topic philosophically, counting on his audience to be unable to tell he is merely pretending. But what of the case where the speaker is simply ignorant? The juxtaposition of affectation and ignorance, coming right on the heels of an account of why pseudo-scientific explanations frequently escape detection, suggests that while the pretentious perhaps deceive only their audience, the ignorant deceive even

[2] Aristotle calls the kind of θεωρία he is about to embark on 'philosophical'; the paraphrase 'scientific' is justified by his description of it as making clear not only the facts but also their causes (see e.g. 1216b32–39).

themselves, as a result of the same superficial understanding of philosophy that makes their audience an easy mark. That is, both types of speaker are merely pretending at philosophy; the difference is that the ignorant pretend unwittingly, moved not by a love of display, but taken in by their own inability to distinguish genuine philosophical discourse from its superficial counterfeit. This suggestion is reinforced by a parallel passage in *Rhetoric* I 2, where we have a similar juxtaposition of causes, and where it is explicit that the effect is a mere pretending at philosophy or science: 'for this reason rhetoric masquerades as political science (ὑποδύεται ὑπὸ τὸ σχῆμα τὸ τῆς πολιτικῆς), and the professors of it as political experts – sometimes from want of education, sometimes from ostentation, sometimes owing to other human causes' (1356a29–30).[3]

Here one might object: even allowing that Aristotle does suggest that honest propounders of empty talk are merely pretending to treat their topics philosophically, still, what could justify him in this? Let it be that they *profess* to treat their topics philosophically, speaking not merely to persuade, but to demonstrate or teach: to prove 'that' by showing 'why' (cf. 1216b26–8, 37–40). Let it be too that they fail to do so – that even when the points they argue for are true, they are not true 'for the reason the argument says' (1217a14–16). That doesn't show that their profession is hollow – that they are merely pretending to do philosophy. You do not as a rule have to hit the mark in order to have been aiming for it. The answer to this objection is that it depends how badly you miss. Different undertakings require different equipment and competencies – not only to succeed, but even to make the attempt. Not having wings (for a start), I cannot even try to fly home from work this evening; because I don't play, I cannot even attempt to perform the Emperor Concerto. To be sure, I do have a (very) superficial understanding of these activities, of what it might look like to engage in them; on this basis I can pretend to attempt them, beating my arms or pounding the keys. But were I actually to carry on in these ways, beating my arms or pounding the keys, I could not be described as *trying* to fly home or to play the Emperor Concerto. At best I would be merely pretending; my 'efforts' would fall too wide of the mark to be genuine attempts.[4]

Though these examples are deliberately fantastic, the point they illustrate is relevant to appreciating how Aristotle thinks about empty talk. Its honest propounders, being 'uneducated' (ἀπαίδευτος), lack certain competencies.

[3] Tr. Roberts. For the imagery see *Grg.* 464b–d.

[4] For rather more nuanced discussion see Albritton 1985, 244–46.

As a result their 'attempts' to treat matters philosophically, to prove 'that' by showing 'why', fall wide of the mark. They fall very wide of the mark: the 'reasons' (λόγοι) they propose are altogether irrelevant, simply 'foreign to' (ἀλλότριος) the subject matter (πρᾶγμα) they purport to explain. This 'failure' is the direct result of their incompetence: 'lack of education' (ἀπαιδευσία) *just is* 'the inability to discriminate' (τὸ μὴ δύνασθαι κρίνειν) between relevant and irrelevant (οἵ τ' οἰκεῖοι λόγοι τοῦ πράγματος καὶ οἱ ἀλλότριοι) (1217a8–10). That's why their theories and arguments are 'empty' (κενός), of a piece with those put forward by pretentious frauds, fit only to be summarily dismissed (cf. *EE* I 1, 1214b28–1215a7). Too wide of the mark to be seriously taken, they are too wide of the mark to be seriously meant; their proponents are merely playing at philosophy and science.

2

I have argued that when Aristotle charges people with 'empty talk', at least sometimes he is charging them with merely pretending at philosophy or science. I have further argued that the charge is grounded in the fact that their theories and arguments fall wildly wide of the mark: so wide as to raise the suspicion that 'the mark' was not their target. (Note that this needn't imply that the accused are knowingly dishonest; often enough, they are simply incompetent – no better than their audiences at telling what sorts of consideration even bear on the topic.) I now turn to consider the underlying causes of this incompetence, as Aristotle understands them. I begin in this section with a passage from *EN* x 9, which shows that chief among these causes is a lack of personal, first hand experience.

Again I begin with some context. Aristotle opens *EN* x 9 by asking whether he has finished what he set out to do; he answers that if theories and arguments (λόγοι) were enough to make men good, then maybe so, but seeing that they're not, we must take up the topic of legislation (1179a33–1180b29). This prompts the question that then occupies the rest of the chapter: where do we learn *that*? Aristotle's answer, roughly and in a word, is nowhere: the science of legislation has yet to be discovered.[5] He arrives at this verdict by a process of elimination. Looking around, there are really only two candidates to choose from: the politicians who practise politics and those sophists who profess to be its teachers; since neither of these are even remotely plausible *teachers* of politics, it must be that the science of legislation awaits discovery.

[5] Stronger: 'our predecessors have left the subject of legislation *unexamined* (ἀνερεύνητον)' (1181b12).

A harsh verdict, to be sure, but of interest here is Aristotle's treatment of the sophists, whom he describes as follows:

> But the professors (οἱ ἐπαγγελλόμενοι) seem very far from teaching it. For on the whole they don't even know what sort of thing it is nor what sorts of things it is concerned with (ὅλως γὰρ οὐδὲ ποῖόν τι ἐστὶν ἢ περὶ ποῖα ἴσασιν); otherwise they would not have classed it as identical with rhetoric or even inferior to it, nor have thought it easy to legislate by collecting laws that are thought well of; they say it is possible to select the best laws, as though even the selection did not demand appreciation and as though right judgment were not the greatest thing, as in matters of music. For while people experienced in any department judge rightly its works, i.e. appreciate by what means or how they are achieved, and which chime with which, the inexperienced must be content if they do not miss entirely whether the work has been made well or ill – as in the case of painting. Now laws are as it were the 'works' of the political art; how then can one learn from them to be a legislator, or judge which are best? It is evident that not even medical men are made from books. Though they try, at any rate, to state not only the treatments, but also how particular classes of people can be cured and should be treated, distinguishing the various habits of body; but while this seems useful to experienced people, to the ignorant it is useless. Surely, then, while collections of laws, and of constitutions also, may be serviceable to those who can see and can judge (δυναμένοις θεωρῆσαι καὶ κρῖναι) what is good or bad and what chimes with what, those who go through such collections without a practised faculty (ἕξις) will not have right judgment (unless it be as a spontaneous gift of nature), though they may perhaps become more appreciative in such matters. (1181a12–b12; tr. Ross, modified)

I begin with a preliminary point, which is that these self-styled teachers of politics are a fair example of the sort of 'empty talker' we have been concerned with. This is not to deny that, in Aristotle's view, their motives are typically impure; no doubt he thinks they are often moved by a love of money or display or 'other human causes' (cf. *Rh.* I 2, 1356a29–30).[6] To allow this is to allow that 'ignorance' is not their *only* problem. But I do insist that ignorance is, not only *among* their problems, but also *the* deficiency Aristotle appeals to in explaining why their profession to teach is so absurd: 'on the whole they do not know even what sort of thing [legislation] is or what sorts it is concerned with' (1181a14). This deficiency is obviously a serious one; though Aristotle does not say as much in so many words, we may expect him to think that it is seldom if ever and even then just by chance that the instruction of these sophists manages even to be

[6] On sophists in Aristotle see Classen 1981.

'relevant', as opposed to 'irrelevant' and 'empty'. In other words, it is very plausible to suppose that Aristotle regards these 'teachers' as incompetent in precisely the way that interests us here.[7]

What then is the cause of this incompetence? Earlier Aristotle observes that these 'teachers' do not *practise* politics (1180b35–1181a1); he also says that some familiarity or experience (συνήθεια, ἐμπειρία), such as is had by those who *do* practise politics, seems necessary (προσδεῖν ἔοικεν) for anyone aspiring to 'know' about politics (περὶ πολιτικῆς εἰδέναι) – that is, for anyone aspiring to a scientific or philosophical understanding (1181a10–12). This suggests that, in his view, the reason these people are incompetent to teach is that they have no experience legislating; put another way, a scientific understanding of politics cannot be acquired without some measure of first-hand experience: as he says, 'not even physicians are made from books'.

Here one might object that the insistence on personal, first-hand experience, of a kind that cannot be got simply from books, stems merely from the fact that politics is a 'productive' science, issuing in certain practical 'works' (ἔργα). But this would be a mistake. To be sure, the mistake is suggested by the analogy with music, which contrasts judging the work as a whole with appreciating the how and the why (1181a19–23); it also sits nicely with the comparison to medicine, in which it is remarked that in medicine even the books are not mere catalogues of treatments, but veritable 'how-to' manuals for treating specific pathologies. Nevertheless, although Aristotle *is* maintaining that the sophists' instruction leaves their students without certain capacities, and that these capacities concern the 'ways and means' of good legislation, still, the capacities Aristotle is talking about are *theoretical* capacities: capacities for discrimination and judgment (κρίνειν). This is clearest from his description of those who *can* profit from books; they are described, not as those who are 'capable of action' (δυνάμενοι πράττειν), but as those who 'can *see* and can *judge*' (δυνάμενοι θεωρῆσαι καὶ κρῖναι) (1181b7–9). Moreover, this emphasis on 'theoretical' capacities is more than sufficient to explain the emphasis on ways and means. This is because a professional interest in the how and the why, far from being the special prerogative of the practical and productive sciences,

[7] This incompetence manifests itself in their supposing that legislation is 'easy', a simple matter of collecting the laws that are well thought of and selecting the ones that are best; to suppose such a thing is to reveal one to be entirely innocent of the real problems in this area, which in Aristotle's view are fundamentally problems of composition, requiring an understanding of 'which [laws] chime with which' (ποῖα ποίοις συνᾴδει) and 'by what means and in what manner' the work as a whole 'is brought to completion' (δι' ὧν ἢ πῶς ἐπιτελεῖται).

is a mark of philosophy and science quite generally and as such: 'we do not think we know a thing (εἰδέναι) until we have grasped the why of it' (*Ph.* II 3, 194b19–20).

I conclude that, in Aristotle's view, the special 'incompetence' we are interested in derives from a lack of personal, first-hand experience. This is a general point, not limited to the productive sciences, as I have just argued and as is confirmed from elsewhere: for example from *EN* II 1, on intellectual virtues quite generally ('intellectual virtue in the main owes both its birth and its growth to teaching, for which reason it requires experience and time' (1103a15–17)), or from *EN* VI 8 (a passage I return to below), which mentions specifically the science of nature (1142a12–17). These passages show that, for Aristotle, experience – not just information or 'data', which can be got quickly from books second-hand, but *experience* – is an essential element in acquiring the competencies characteristic of the true scientist, even in 'speculative' disciplines that are neither practical nor productive.

3

In the passages we have been considering, we see Aristotle complaining of a certain form of intellectual malpractice. I have argued that this malpractice consists in a kind of pretending at philosophy or science and manifests a certain incompetence for genuine philosophical or scientific work. I have further argued that, in Aristotle's view, this incompetence may be traced to a lack of personal, first-hand experience. I now propose to look closer at Aristotle's understanding of the effects of this kind of experience on our intellectual make-up. I argue that he thinks the lack of it leaves us intellectually 'young' or 'immature'. One consequence of this immaturity is that very often (though unbeknownst to ourselves) our talk is 'empty' – that is, we make assertions without conviction in the truth of what we say.

My first point is that, in Aristotle's view, experience of the kind in question takes *time* – enough time to put it out of reach of the young (νέος δ' ἔμπειρος οὐκ ἔστιν· πλῆθος γὰρ χρόνου ποιεῖ τὴν ἐμπειρίαν) (*EN* VI 8, 1142a15–16). The reason is not just that there is a lot to learn (though there is that), but also and crucially that learning is (in part) becoming convinced, and *that* takes time. This, I submit, is the idea behind his well-known remark that, in scientific matters, the young speak without conviction (οὐ πιστεύουσιν οἱ νέοι ἀλλὰ λέγουσιν) (*EN* VI 8, 1142a16–20). His point is not that there is a limit to how much young people can 'get up', well enough to recite back, but rather that, when they do speak, they do so without being convinced of, without *relying* on and *trusting* in, the truth of what they say.

To illustrate, doubtless many young people have heard it said (e.g.) that riches are hard to acquire but easy to lose, or that most things end badly, or that most people can't be trusted (examples from *Rh.* II 13). Yet as a rule, Aristotle thinks, these dreary sayings make no impression on the young; *that* requires repeated first-hand experience. That's why, he says, young people are not greedy; not because no one has told them that money is easy come, easy go, but because 'they have not yet made trial of want' (διὰ τὸ μήπω ἐνδείας πεπειρᾶσθαι). Similarly, he says, the young 'think no evil, but believe in goodness, because as yet they have not seen much wickedness' (διὰ τὸ μήπω τεθεωρηκέναι πολλὰς πονηρίας) – that is, seen it first-hand, in those they know personally. Young people are gullible, he says, 'because as yet they have not often been deceived' (διὰ τὸ μήπω πολλὰ ἐξηπατῆσθαι); they are bullish, not just 'because they are heated, as with wine', but also 'because they have not had many disappointments' (διὰ τὸ μὴ πολλὰ ἀποτετυχηκέναι) (*Rh.* II 12). In a word, young people are unmoved by such sayings because for them they are *just words*, out of touch with the reality they know first-hand (primarily themselves, secondarily their friends). It is not that the young never assert such things, or that when they do so they are lying (deliberately deceitful). It's rather that in their mouths these sayings are just talk: they speak without being convinced (οὐ πιστεύουσιν οἱ νέοι ἀλλὰ λέγουσιν). This is because it takes time for these 'lessons' to become things they *trust*, things they *rely on*: genuine parts of their intellectual make-up, canons by which they tell true from false. As he says in EN VII 3, 'those who have just begun to learn a science (οἱ πρῶτον μαθόντες) can string together its phrases, but do not yet know it; for *it has to become part of themselves* (δεῖ γὰρ συμφυῆναι), and that takes time' (1147a21–22).

The suggestion, then, is that for Aristotle, at least one thing experience contributes to knowledge is conviction or trust (πίστις). In his view, when we experience things repeatedly first-hand, we 'make trial' (πεῖραν λαβεῖν) of them; such trials earn our trust, not in the sense of justifying it (though they may do that too), but in the sense of evoking or effecting it. This way of thinking is illustrated especially clearly in a passage from the *EE*, about the unique stability of the primary form of friendship:

> There is no stable friendship without trust (πίστις), nor trust without time; for it is necessary to make trial, as Theognis says: 'You cannot know the mind of man nor woman | Before have you tried them as you try cattle.' (*EE* VII 2, 1237b12–16; tr. Rackham, modified)

Though the comparison is unsettling, this passage is illuminating for its succinct description of the effects that 'experience' of things – of *having made trial* of them – has on our intellectual character. When we have repeatedly made 'trial' of a thing we are caused to rely on it; as a consequence, when we speak from experience, we speak with conviction.

Aristotle's view of the primary form of friendship is relevant here in another way. If these friendships involve trust, and therefore take time, it follows that they can't be formed quickly. Now, we might worry that this implication is refuted by facts; Aristotle, however, thinks otherwise, and his explanation is interesting:

> Nor is one a friend without time, but they only wish to be friends, and such a condition very easily passes for friendship (καὶ μάλιστα λανθάνει ἡ τοιαύτη ἕξις ὡς φιλία). For, when they are eager to be friends, they render one another all friendly services, and because of that they think, not that they want to be friends, but that they really are friends. But in fact, as in everything else, so too in friendship; it is not that, if people want to be healthy, they are healthy, so that neither, even if they want to be friends, are they actually friends already. Here's a proof: those who without trial (ἄνευ πείρας) are disposed in this way – i.e. want to be friends – are easy to malign; for though men are not easy to malign where they have tried one another (πεῖραν δεδώκασιν ἀλλήλοις), where they have not they will be convinced whenever the maligners proffer tokens in guarantee (πεισθεῖεν ἂν ὅταν σύμβολα λέγωσιν οἱ διαβάλλοντες). (1237b16–27)

The parallels implicit in this passage, between playing at friendship and playing at philosophy, are striking. We are not authorities about who our friends are; wanting to be someone's friend doesn't make us so; in fact we are often deceived, and not just about whom we *may* trust or rely on, but about whom we *do* trust and rely on. This deception can be revealed, objectively, our protests to the contrary notwithstanding. False trust is fickle and easily shaken; regardless of what we may want or say, we hardly *do* trust those we can be brought to think ill of so easily. As personal immaturity leaves us an easy mark for slanderers, so too intellectual immaturity (ἀπαιδευσία) leaves us an easy mark for pseudo-science and philosophy.

We were asking about the impact of personal experience on the development of intellectual character. The passages we have been considering suggest that, in Aristotle's view, the lack of such experience leaves us arrested in a state of intellectual immaturity. One manifestation of this immaturity is a readiness to speak without conviction, not with an intention to deceive others, but deceived ourselves about where our trust – and consequently our 'friendship' or 'regard' – really lies.

4

I now turn to Aristotle's scientific writings, in order to illustrate how frequently he complains there of the phenomenon we have been exploring ('empty talk'). As before, my interest is not in the vocabulary of 'emptiness' per se, but in a pathology that Aristotle sometimes uses that vocabulary to describe: a kind of playing at natural science, propounding theories and arguments without conviction in or regard for their truth.

The first passage I want to consider is taken from *Generation and Corruption* I 8. Aristotle has been discussing certain theories of 'action and passion'; having singled out Leucippus and Democritus for special praise, he goes on to contrast them with the Eleatics, who (to judge from what they say) don't even believe in action and passion, maintaining instead that reality is 'one' and 'immobile'. But they are not to be believed:

> Although these results seem to follow in respect of the arguments (ἐπὶ τῶν λόγων), it is next to madness to judge (δοξάζειν) that they are so in respect of the things themselves (ἐπὶ τῶν πραγμάτων) (for no lunatic is so far gone as to think (δοκεῖν) that fire and ice are one, but only what is admirable and what customarily seems so – there are some who on account of madness think *these* are no different). (325a19–25)[8]

In this passage Aristotle is charging the Eleatics with not believing what they say. Yes, he does allow that they give reasons for the views they profess, and that their reasoning seems to have something to it.[9] But what does he think it would be, not just to profess, but to *judge* (δοξάζειν) that reality (τὰ πράγματα) is as the Eleatics say it is? Near lunacy, or rather beyond lunacy – even those who are literally mad do not go so far as to think fire and ice are one. What then are we to conclude? Surely not that the Eleatics are beyond literally mad, nor that they are knowing charlatans or frauds – deliberately dishonest in working out their philosophy. The alternative is that they do not mean what they say; caught up in the validity of their 'reasonings' (λόγοι), they lose sight of the fact that they are professing to decide about *things* (πράγματα).

Aristotle thinks this phenomenon is depressingly common; indeed, he thinks it is an occupational hazard of scientific work. This is clear from the next passage I want to consider, which is taken from *de Caelo* II 13. In this chapter Aristotle has been considering various theories about the support

[8] Text following Rashed (2005), see his note on the passage on pp.138–39.

[9] '[T]hey give reasons for the views they profess': οὕτως καὶ διὰ ταύτας τὰς αἰτίας ἀπεφήναντο περὶ τῆς ἀληθείας (325a16–17).

of the earth; though he does not think that these theorists get close to the bottom of the matter, his explanation of why they are content with their theories, despite their manifest inadequacy, is unusually sympathetic:

> They seem to inquire to a point, but not so far as the problem admits. *It is what we are all accustomed to do* (πᾶσι γὰρ ἡμῖν τοῦτο σύνηθες), to direct our inquiry not to the matter itself (πρὸς τὸ πρᾶγμα), but to the views of an opponent (πρὸς τὸν τἀναντία λέγοντα); even when inquiring by oneself (αὐτὸς ἐν αὑτῷ), one inquires only until one is out of objections (μέχρι περ ἂν οὗ μηκέτι ἔχῃ ἀντιλέγειν αὐτὸς αὑτῷ). That's why one who intends to inquire well must be ready in bringing forward objections proper to the genus; this comes of having studied all the differences (ἐκ τοῦ πάσας τεθεωρηκέναι τὰς διαφοράς). (294b6–13)

This passage makes clear that, in Aristotle's view, inquiring with regard to an opponent, as opposed to the subject at hand, is not the exception but the *rule*: πᾶσι γὰρ ἡμῖν τοῦτο σύνηθες.[10] Indeed, it is so much the rule that the proper remedy is not to repress the instinct (impossible), but precisely to harness it, intensifying our contentiousness, making ourselves downright quarrelsome (ἐνστατικός), albeit in a very particular way, namely, by arming ourselves so thoroughly with the salient facts that, in our case, inquiring with regard to an opponent *just will be* inquiring with regard to the subject at hand.[11] This shows that, in his view, taking one's eye off the subject is not something we have to be especially perverse to fall into, but something we must take special pains to avoid. When we do not, though we may not be aware of it, we speak without conviction – without conviction, that is, about the reality (πρᾶγμα) our discourse is ostensibly about. *That's* how it happens that people aver (e.g.) that the earth floats on water. They say such things, not intending to deceive, but because they are not thinking; or rather, they are thinking, but not about the subject they're talking about, being focused instead on certain 'reasonings' (τοὺς ἐναντίους λόγους).[12]

10 Cf. *Grg.* 457d–458b, where we find the idea of inquiring 'in relation to the subject at hand' (πρὸς τὸ προκείμενον ἐν τῷ λόγῳ), and of being ambitious for victory in speaking 'in relation to the thing' (πρὸς τὸ πρᾶγμα), in both cases as opposed to 'in relation to *you*' (πρὸς σέ), sc. one's interlocutor. Also 453c:

> Now why is it that, having a suspicion of my own, I am going to ask you this, instead of stating it myself? It is not on your account, but with a view to the argument, that it may progress *as may best reveal to us the point we are discussing* (οὐ σοῦ ἕνεκα, ἀλλὰ τοῦ λόγου, ἵνα οὕτω προΐῃ, ὡς μάλιστ' ἂν ἡμῖν καταφανὲς ποιοῖ περὶ ὅτου λέγεται).

11 Cf. *EE* I 3, 1214b28–1215a7.

12 Cf. *SE* 1, 156a6–10:

> Since it is impossible in a discussion to bring in the actual things discussed, we use their names as symbols instead of them (τοῖς ὀνόμασιν ἀντὶ τῶν πραγμάτων χρώμεθα ὡς συμβόλοις), and therefore we suppose that what follows in the names, follows in the things as well (τὸ συμβαῖνον ἐπὶ τῶν ὀνομάτων καὶ ἐπὶ τῶν πραγμάτων ἡγούμεθα συμβαίνειν), just as people who calculate suppose in regard to their counters.

This danger, mitigated by experience, is exacerbated by poor education, which makes one 'facile' and 'inconsiderate of facts'. Consider the following passage from *GC* I 2. Here Aristotle has been discussing the superiority of Democritus' physical theory to that found in the *Timaeus*. On setting the *Timaeus* aside he offers the following concluding reflection:

> Inexperience is responsible for the diminished capacity for seeing the admitted facts together (αἴτιον τοῦ ἐπ' ἔλαττον δύνασθαι συνορᾶν τὰ ὁμολογούμενα). That's why those who have lived more among natural phenomena (ἐνῳκήκασι μᾶλλον ἐν τοῖς φυσικοῖς) are better able to lay down principles capable of connecting a broad range of things (αἳ ἐπὶ πολὺ δύνανται συνείρειν), while those who from many reasonings have become inconsiderate (ἀθεώρητοι) of facts, having looked at but few, make pronouncements easily (ἀποφαίνονται ῥᾷον). (316a5–10)

This passage describes the ill-effects of a certain kind of quasi-philosophical or quasi-scientific education, one focused above all on 'theories' or 'reasons' or 'arguments' (λόγοι). The effect of this 'education' is to make us narrow, facile, and inconsiderate of facts (316a8–10). We might say of its graduates, interpreting their pronouncements *au pied de la lettre* – that is, as about some department of reality – that unbeknownst to themselves, they speak without conviction. Preoccupied with λόγοι, they fail to notice that they are not even thinking about reality (πράγματα, ἔργα). Like the φιλόδοξοι of Plato's *Republic*, their intellectual work is not disciplined by a love for the sight of the truth; that is not what their 'reasoning' and discourse is 'towards' (πρός).

If this seems harsh, Aristotle himself puts just this construal on proponents of the Timaean theory in *Cael* III 7 (the last passage I consider in this section). This theory, he says, requires its proponents to assert (as they do assert) that not all the elements can be generated out of one another. But this, he complains, is 'absurd' (ἄτοπον):

> For neither is it reasonable (εὔλογον) that one alone should have no part in the transformations, nor is it evident to the senses, but rather it is evident that all alike change into one another. The result is that, speaking *about* the observed facts, they say *what doesn't agree with* the observed facts (συμβαίνει δὲ περὶ τῶν φαινομένων λέγουσι μὴ ὁμολογούμενα λέγειν τοῖς φαινομένοις). (306a3–7)

Having described this result in a way suggestive of paradox, he proceeds to offer the following diagnosis:

> The cause of this is that their first principles are not well taken, but instead they want to trace everything back to certain definite *opinions* (πάντα βούλεσθαι πρός τινας δόξας ὡρισμένας ἀνάγειν)... *[O]wing to their love*

> *of these opinions* (διὰ τὴν τούτων φιλίαν), they do the same as those defending theses in discussions; they abide any result, as though they were holding true principles – as if some principles did not need to be judged from their results (κρίνειν ἐκ τῶν ἀποβαινόντων), and especially their τέλος. And while the τέλος of productive knowledge is the work (τὸ ἔργον), that of natural science is what appears to the senses always and decisively (τὸ φαινόμενον ἀεὶ κυρίως κατὰ τὴν αἴσθησιν).[13] (306a7–17)

The charge is unmistakable: these theorists 'inquire' like someone defending a thesis – for the sake of argument, without regard for its truth. Especially arresting is the language of 'love' (φιλία) and 'opinion' (δόξα). These theorists do have an affection and regard for something: certain definite 'opinions', which serve them in their 'investigations' as first principles. The trouble is that their professed topic is *not* the consequences of those principles, but rather the observed facts; these, as they themselves would admit, are what it is precisely the point (τέλος) of their undertaking to understand and explain. Interpreted this way, as contradicting the facts they are speaking about, we can only conclude that they don't mean what they say – like one defending a thesis, they speak without conviction (οὐ πιστεύουσιν ἀλλὰ λέγουσιν).

These passages show that Aristotle thinks 'empty talk' is very common – not just in practical matters, but also in natural-scientific ones. This is not some bizarre perversion, which we would have to be unusually corrupt to fall into. On the contrary, he thinks, it is a standing risk in many kinds of intellectual work, which we must take special pains to avoid. Though he allows that reasoned argument is in a way the very badge of scientific and philosophical inquiry, he nonetheless believes that developing a facility in argument will not by itself develop a regard for and confidence in the realities (πράγματα) this inquiry is ultimately about. Regard for and reliance on facts is an acquired habit, developed over time and by trial. When we are deficient in this habit, and grow accustomed to arguing in a way that is facile and reckless, without concern for the facts we are ostensibly reasoning about, then whether we realize it or not, we stand fairly accused of but playing at philosophy and science, holding forth, to be sure, but without regard for or conviction in the truth of our pronouncements.

[13] Taking the adverbs ἀεὶ and κυρίως as qualifying the participle φαινόμενον, rather than the relationship of the predicate (τὸ φαινόμενον . . . κατὰ τὴν αἴσθησιν) to the subject (τέλος, carried over from the previous clause): that is, 'the τέλος is what always and decisively appears', rather than 'the τέλος is, always and decisively, what appears'.

5

The passages we have just been considering are taken from texts given to Aristotle's astronomy and 'chemistry' – not his strongest suits. Indeed, Aristotle's failings in these areas might lead us to wonder whether his remarks in these passages are only so much invective, from which it would be a mistake to draw serious morals.[14] With this in mind I turn briefly to consider some passages from his biological writings, in which he is clearly holding a better hand. My idea is that in such cases it will be easier to sympathize with his complaints (and thus to take them seriously). The passages I want to consider are all taken from *GA* III–IV. Aristotle is very dismissive of the views he considers, and much of the language he uses, both in describing the views and in diagnosing the 'failings' of their proponents, will be familiar from passages we have already considered. I discuss three examples and then draw some conclusions.

First, against the view that (Selachia apart) all fish are female, the so-called 'males' only differing from the females in not reproducing, Aristotle complains as follows (*GA* III 5, 8). First, the 'so-called' males *do* differ from the females in other ways; they lack a uterus and possess the organs for milt (indeed, when the season is right for it, semen *is seen* oozing out of them). Second, the 'difficulty' that leads to this view is no difficulty at all, 'being easy to solve for those who listen to facts' (εὔλυτος τὸ συμβαῖνον ἀκούσασιν)' (755b22–23). Though it is true that, unlike other animals that divide the sexes, oviparous fish produce a vast number of eggs, this only seems to present difficulty because the theorists in question haven't 'connected' (συνεωράκεσαν) it with the fact that, unlike those other animals, fish eggs finish developing outside the animal. Indeed, we find the same combination in cephalopods and crustaceans, which *are seen* copulating. The reason the proponents of this theory have gone so badly astray (αἴτιον . . . τῆς ἀγνοίας) is that although the differences in how animals copulate and reproduce, being manifold, are unclear, these theorists, 'basing their study on a few cases, think the same must hold good in all' (756a4–5).

14 This is a concern Aristotle is alive to: cf. the opening of *Resp.* (470b5–12):

> Regarding respiration, though a few previous physicists have discussed it, when it comes to *why* animals respire, some made no pronouncement at all (οὐδὲν ἀπεφήναντο), while the rest, though they did discuss it, discussed it poorly, in a manner quite unfamiliar with the facts (ἀπειροτέρως τῶν συμβαινόντων). Further they say that all animals breathe, but this is not true. Thus we must go through these points first, *lest we seem to be making empty accusations against the departed* (ἀπόντων κενὴν κατηγορεῖν).

Second, against the view that female fish conceive from swallowing the seed: first, the correct view – that on being laid by the females, the eggs are fertilized by milt shed on them by the males – chimes with other facts, for example that milt and eggs both develop and are emitted at the same time as one another (756a8–14). Second, this view 'only makes sense' (συμβαίνει κατὰ λόγον) (756a14–15), viz. when you consider how it is with other oviparous animals: roughly speaking, most eggs get fertilized and no fetation without fertilization (*modulo* differences stemming from whether the eggs in question are perfect or imperfect) (756a15–30). Indeed, in view of these facts, the problem rather becomes to explain how these other theorists went wrong. One factor is that certain fish 'copulate' so quickly that the fact passes unnoticed, even by fishermen (none of *them* watches for this sort of thing for knowledge's sake – though the thing has in fact been seen) (756a30–34). That's why, not seeing them copulate, but seeing them instead swallowing milt and eggs, even fishermen repeat 'the same simple-minded tale' (τὸν εὐήθη λόγον) (756b5–6), 'babbling the same drivel as even Herodotus the storyteller' (καὶ τεθρυλημένον ὅνπερ καὶ Ἡρόδοτος ὁ μυθολόγος) (756b6–7), 'not considering that this is impossible' (οὐ συνορῶντες ὅτι τοῦτ' ἔστιν ἀδύνατον) (756b3–8) – impossible because, once swallowed, the milt must end up in the intestines, while the eggs are in the uterus (756b8–12). A similar story is told of birds and even some quadrupeds (e.g. weasels): the proponents of such stories speak 'without consideration' (ἀσκέπτως) and 'with excessive simplicity' (λίαν ἁπλῶς) (756b17–18). They are 'deceived by reasoning' (ἐκ συλλογισμοῦ διαψευδόμενοι) (756b18–19), their mistake being like that made about fish (certain birds are not often seen copulating, but are often seen locking beaks – but it is 'strange' (ἄτοπον) that these theorists 'do not reckon' (μὴ συλλογίζεσθαι) how the semen gets from intestine to uterus (756b27–29).[15]

Third, against Empedocles' view that sexual differentiation occurs in the mother (if the uterus is hot, we get a male, if cold, a female), Aristotle complains that it is too facile (ῥᾳθυμότερον) just to suppose (οἰόμενος) that male and female embryos differ only by hot and cold, when you can

[15] 'Also too deceived (λίαν διεψευσμένοι)', Aristotle says, 'are those who make a foolish statement (εὐηθικῶς . . . λέγοντες) about the trochus and the hyena' (757a2–3). The statement is that these animals have the sexual organs of both males and females, the trochus mounting itself, the hyena taking turns by season, mounting and being mounted. Aristotle thinks this ridiculous, among other reasons because it is plainly refuted by observations that are easy to make (ὦπται γὰρ ἡ ὕαινα ἓν ἔχουσα αἰδοῖον· ἐν ἐνίοις γὰρ τόποις οὐ σπάνις τῆς θεωρίας) (757a7–8). Here again, though there are observations to support these theories – for example 'hyenas have under the tail a line like the pudendum of the female' – the 'observations' were made carelessly (διὸ τοῖς ἐκ παρόδου θεωροῦσι ταύτην ἐποίησε τὴν δόξαν) (757a11–12).

actually see (ὁρῶν) great differences in their respective parts, for example their sexual organs (*GA* IV 1, 764a13–15). Nor does Empedocles even look into or try to state the difference in how the male and female parts actually develop (764a12–23). Or, if his theory is that hot and cold cause this too – not just the difference in sex of the embryos as a whole, but also their different sexual parts – he at least ought to have said so, this being virtually the whole problem in this area (764a25–26). And no small job it would be (οὐ μικρὸν δὲ ἔργον), starting with just hot and cold and ending with the presence or absence of a uterus or the parts used in intercourse (764a26–32). Again, male and female are often found together in the same part of the uterus ('this we have sufficiently observed (ἱκανῶς τεθεωρήκαμεν) by dissection in all the vivipara, both land animals and fish' (764a3336)). Aristotle remarks:

> if he had not seen this too (συνεωράκει), it was only reasonable for him to miss in assigning this cause (εὐλόγως ἡμάρτανε ταύτην τὴν αἰτίαν εἰπών); but if he had seen it (ἑωρακώς), it is strange (ἄτοπον) that he should still think the heat or cold of the uterus to be the cause, since on his theory both these twins would have become either male or female, but as it is we do not see this to be the fact. (764a36–b3)

And as for his view that 'the parts of the embryo are sundered, some being in the male and some in the female parent, which is why they desire intercourse with one another' – this is not even worth discussing (τάχ' ἂν εἴη πολλὰ λέγειν), being sheer invention (ὅλως γὰρ ἔοικεν ὁ τρόπος τῆς αἰτίας πλασματώδης εἶναι) (764b8–10).[16]

With all these theories it is the same: though they do have something to be said for them, their proponents are in various ways inept – facile, narrow, incurious, under-informed, well-informed but not in the habit of *connecting* the pieces of information they have, seemingly unaware what considerations are relevant (is there a way from the mouth to the uterus?), or of what it even would be to explain certain phenomena (e.g. the differentiation of the sexes, that it would be to explain the formation of male and female reproductive organs), in some cases repeating tall tales, in others simply making things up. These are the sorts of consideration Aristotle appeals to in explaining why these theorists are content with – content to the point of propounding – theories and arguments that are manifestly no

[16] Or again, against those who maintain that if a man copulates with one of his testes tied up, the child will accordingly be male or female. The proponents of this view 'are prophesying what will happen from likelihoods' (μαντευόμενοι τὸ συμβησόμενον ἐκ τῶν εἰκότων), 'jumping to the conclusion that it is so before seeing that it is so' (προλαμβάνοντες ὡς οὕτως ἔχον πρὶν γιγνόμενον οὕτως ἰδεῖν) (765a26–29).

good. It is not that they are duplicitous or mad, but rather that they are intellectually 'young' or 'immature'; lacking the competence to do more than play at philosophy or science, they slip unawares into empty talk, proffering theories and arguments without conviction.

* * * * *

I have argued that Aristotle charges his predecessors in philosophy and science with a certain kind of incompetence; this incompetence reflects a superficial understanding of what philosophy and science are, and leaves them incapable of even attempting scientific or philosophical work – the most they can do is play or pretend. I have also argued that though this verdict is harsh, it's not an empty slur; it's a theme in Aristotle's writings and it's theorized. A little more concretely, the incompetence in question is a form of intellectual immaturity, stemming from a lack of experience and tantamount to a lack of intellectual seriousness; as such it manifests itself in the propounding of theories and arguments without regard for their truth. Moreover, though it is an aberration, it is not an uncommon one; in Aristotle's view, it is a standing risk in virtually all scientific or philosophical work. Further, he thinks, this risk is often exacerbated by an 'education' that over-emphasizes argumentation (λόγοι) at the expense of a broad, deep, and personal familiarity with the realities (πράγματα) that *true* reasoning is disciplined thinking about. The result is an intellectual 'youthfulness' in which facts, even when they are mentioned, are not relied upon; that is, they do not function in our intellectual economy in the way that they do function – namely, as canons for discriminating relevant from irrelevant and true from false – when our reasoning is aimed at truth.

CHAPTER 12

The scientific role of eulogos *in Aristotle's* Cael *II 12*

Andrea Falcon and Mariska Leunissen

In the natural treatises, especially when empirical evidence is sparse or non-existent, Aristotle sometimes appeals to what is '(un)reasonable' (*eulogos*, οὐκ ἀλόως, *alogos*) or to what is 'against expectation' (*paralogos*) in order to settle puzzles or even to generate scientific explanations. Robert Bolton recently suggested that the presence of these terms – especially in *On the Heavens* (*Cael*), but also in other natural treatises – typically signifies that in these cases Aristotle has moved away from the natural scientific mode of investigation (i.e., from investigating things *physikôs* and 'in accordance with perception' or *kata tên aisthêsin*) and is using the dialectic method instead (i.e., investigating things *logikôs* and 'in accordance with reason' or *kata ton logon*, by starting from reputable opinions), and that his explanations constitute merely dialectical solutions rather than natural scientific knowledge.[1]

As we will make clear below, however, nowhere in *On the Heavens* does Aristotle actually use dialectic as outlined in the *Topics*, and he is, moreover, quite explicit about his intentions to provide natural scientific explanations while at the same time appealing to what is *eulogos*.[2] As we will argue in this chapter, Aristotle's appeal to what is *eulogos* or not in natural inquiry should not be interpreted as *ipso facto* indicative of a dialectical method, but rather can also be understood as a particular application of his natural scientific method that is appropriate for cases where empirical evidence

The idea of this chapter came to us at the annual meeting of the European Society of Ancient Philosophy held in Athens, 29–30 March 2012, where we separately presented on different sections of *Cael* II 12. Drafts of this chapter were presented at two Aristotle workshops held in May 2013, the first at the University of Leiden and the second at Rutgers University. We are grateful to all those who participated in the discussions at these workshops for their helpful feedback. Special thanks go to David Charles and Robert Bolton, who generously read and commented on earlier versions. We would also like to thank David Ebrey who has given us feedback on the penultimate draft of the chapter.

[1] Bolton (2009, 51–82).

[2] By 'scientific explanation' we mean an explanation that is *in principle* answerable to empirical evidence.

is lacking. On our reading, Aristotle generates explanations in terms of what is *eulogos* – in *On the Heavens*, but also elsewhere whenever a given natural scientific domain is empirically underdetermined – by showing how the case at hand makes very good sense (or not) given some relevant background principles or hypotheses that have already been established, usually in an empirical manner. By establishing that something is *what one could reasonably expect to be the case* in light of what is already known to be the case about that domain, Aristotle is not merely offering dialectical solutions, nor is he providing us with natural scientific knowledge in the strict sense. Rather, he is able – at a minimum – to reduce our puzzlement about these phenomena, and – at a maximum – to offer 'convictions (*pisteis*) suitable for humankind',[3] until, perhaps, more progress can be made and hitherto unknown empirical evidence can be uncovered.

Furthermore, we see no conflict between appeals to what is *eulogos* in contexts where Aristotle is engaged with views of his predecessors, or where he is providing *logikôs* or (otherwise) dialectical arguments, and his overarching aims to generate scientific explanations to the extent the available empirical evidence allows him to do that. We believe that Aristotle's methods in the natural treatises are flexible and manifold, and that there is no reason why he cannot make use of *both* his dialectic *and* his scientific methods within one treatise or even one discussion. Thus, our chapter also offers an alternative picture to the tendency in the scholarship on Aristotle to characterize his natural treatises as being either entirely dialectical in method or as simply following the (exact same) scientific methods introduced in the *Posterior Analytics*.[4]

We develop our account on the basis of a close reading of what is arguably the most interesting text in the corpus, especially for its methodological implications: *Cael* II 12. First of all, however, we would like to offer a general analysis of the ways in which Aristotle appeals to what is *eulogos* in the natural treatises and their connections to his dialectical and scientific methods.

[3] *Cael* II 5, 287b28–288a3.

[4] This tendency can be traced back to G. E. L. Owen (1986), 151–64, who credits Aristotle with a clear distinction *within natural philosophy* of areas of inquiry that are empirical and areas of inquiry that are dialectical. According to Bolton, whose work on dialectic is in part a response to Owen's remarks, if natural philosophy follows the general theory of science outlined in the *Posterior Analytics*, then its starting points and controlling factors must be data that are derived from experience. Hence, when explanations cannot be confirmed by empirical data, they are not warranted for acceptance as natural scientific knowledge (*epistêmê physikê*). It is the combination of these two claims that leads Bolton to the conclusion that there are two mutually exclusive standards of inquiry used to reached warranted results in the study of nature, namely a scientific and a dialectical one.

Dialectic, science, and Aristotle's use of *eulogos* in natural scientific inquiry

Methodological peculiarities in Cael *II*

In *Cael* II Aristotle discusses the features and motions of the heavens as a whole and of the individual heavenly bodies (including the earth, even though it is not technically speaking a heavenly body). And in three cases, in explaining why the heavenly bodies have the particular attributes they do, Aristotle follows a very specific and notable pattern: he first provides a methodological comment about the limits of the empirical evidence that can be gained about the features and motions of the heavenly bodies, then makes clear that he nevertheless intends to offer some kind of explanation, and finally provides a – somewhat peculiar – teleological explanation derived from a teleological principle.[5]

Two additional features of his explanatory strategy stand out. First, the explanations in *Cael* II 5 and 12 are explicitly characterized as having a lesser status. Aristotle states that they (merely) reflect 'what appears to be the case' (*Cael* II 5, 288a1–2 and II 12, 291b25: τὸ φαινόμενον), that they do not involve 'more accurate necessities' (*Cael* II 5, 287b34: τὰς . . . ἀκριβεστέρας ἀνάγκας), and that they provide 'small solutions' (*Cael* II 12, 291b27: μικρὰς εὐπορίας). Second, these explanations are introduced as accounts in virtue of which certain *aporiai* are solved (*Cael* II 5, 288a8–9), or at least will no longer appear puzzling or *alogos* (*Cael* II 12, 292a18–19), and because of which the data at our disposal will no longer appear *paralogon* (*Cael* II 12, 292a21–22). Both of these features of Aristotle's explanatory strategy raise questions about the scientific status of these explanations.

Below (in the section on *eulogos* in *Cael* II 12, pp. 230ff.) we will argue (i) that Aristotle's methodological remarks in *Cael* II 12 are serious and fitting with remarks about the study of the heavens Aristotle makes elsewhere; (ii) that his explanations are scientific, albeit of a weaker, less necessary kind; and (iii) that his appeals to what is *eulogos* are part of his *scientific* strategy in generating such explanations when empirical data are lacking or inconclusive. Before turning to this specific text, however, let us first make clear what, in general and in outline, we take to be the role of dialectic in scientific investigations and how Aristotle's appeals to what is *eulogos* in the natural treatises are related (or not) to his dialectic and scientific methods.

[5] See *Cael* II 3, 286a3–9; II 5, 287b28–288a3; and II 12, 291b24–28 with 292a14–22. On the teleological explanations offered in these chapters, see Leunissen (2010, 152–75).

A sketch of the role of dialectic in scientific investigations and its application to Cael *II 12*

As Aristotle makes clear in the *Topics*, there are three uses of dialectic, which he broadly defines as a deductive mode of reasoning that starts from reputable opinions: it can be used in intellectual training, in ordinary discussions, and 'in relation to the philosophical sciences'.[6] Aristotle's characterization of the last use is worth quoting in full, as it illustrates how dialectic is used in the sciences:

> [It is useful] for the philosophical sciences, because by being able to puzzle on both sides of something we will perceive with more ease what is true and what is false in every matter. In addition, [it is useful] for the first [principles] regarding each of the sciences. For it is impossible to say something about them on the basis of the ones [i.e., principles] that are proper to the science at hand, since the principles are prior to everything, and it is necessary to go through them via the reputable opinions about them. And this is what is proper or most fitting to dialectic.[7]

Based on this passage, we can formulate two possible uses of dialectic in Aristotle's natural science. The first use is this: if we grant that natural science is the kind of theoretical science Aristotle has in mind when mentioning the philosophical sciences, then dialectic helps the natural scientist by enabling him to work his way through the puzzles pertaining to some issue in the appropriate way and thereby to '*perceive* with more ease what is true and false in every matter'. In this case, dialectic does not help natural scientists *establish* the truth concerning something, since that is the exclusive role of natural science itself; rather, it helps them see why the alternatives they rejected as false were indeed false, and why the views they accepted as true were indeed true:[8] dialectic perhaps enables one to become immune to being persuaded out of knowing something scientifically. This role of dialectic in science as described here, then, is mostly *post hoc*: it is only *after* scientists have provided their explanations (or candidate scientific demonstrations) or identified their principles that dialectic is required to come in to 'clarify' the outcomes of their investigations,[9] although it can, of course, also be used during those investigations *to convince others*.[10]

Let us turn to the second possible use of dialectic. Our passage says that dialectic is used most properly in connection with the first principles of a science, which presumably involves the case of someone challenging those

[6] *Top* I 2, 101a26–27. [7] *Top* I 2, 101a34–b3. [8] Cf. *NE* VII 14, 1154a24–25 and *APo* I 2, 72a37–b2.
[9] See e.g. *NE* VII 1, 1145b2–7. [10] See e.g. *Cael* I 10, 279b4–12.

principles. Since this person cannot be refuted from within the science (as this person is challenging the most fundamental principles of that science), dialectic can be employed as a way to establish the principles *in an alternative way*. This use of dialectic applies to natural science, as is evidenced in the following passage:

> To investigate whether what exists is one and motionless does not belong to the investigation about nature. For just as there is nothing to say for the geometer to the person who denies his principles, but rather it belongs to a different science or to a science common to all, so too there is nothing [to say] for the [natural scientist] about his principles. For there would no longer be a principle, if [what exists] were one alone and one in that way, for a principle is of something or of some things. The investigation of whether [what exists] is one in that way would be similar to arguing against any other posit that is said for the sake of argument (πρὸς ἄλλην θέσιν ὁποιανοῦν διαλέγεσθαι τῶν λόγου ἕνεκα λεγομένων) . . . or to refuting an eristic argument (ἢ λύειν λόγον ἐριστικόν) . . . We must assume that the things that exist by nature are – either all or some of them – in motion. And it is clear from induction.[11]

The assumption that things that exist by nature are in motion is a first principle of Aristotle's natural science, which means that someone who challenges that assumption cannot be refuted by the natural scientist speaking as a natural scientist, since the sceptic, after all, denies the very foundation of every claim to knowledge the former can make.[12] Here dialectic provides a way to defend the first principles of a science *from outside the boundaries of that science*.

In order for *Cael* II 12 to count as dialectical in method, then, it should preferably exhibit one of these two uses. But as will become clear in the second section below (pp. 230ff.), Aristotle is not defending any first principles of his natural science in *Cael* II 12,[13] nor is he trying to gain a stronger conviction for himself or others by working through the discussion of *aporiai* that arise given either the truth or falsity of his explanations. So the chapter could be dialectic only in a loose sense, either because it would generate accounts by appealing not to empirical evidence but rather to reputable opinions, or because the arguments would be a priori and developed independently of any empirical evidence. Although we believe that dialectic in this looser sense can be used in natural inquiry, the specific *aporiai* discussed in *Cael* II 12 are not inherited from the philosophical

11 *Ph* I 2, 184b25–185a14. 12 Cf. *Cael* III 1, 298b14–18.

13 Bolton (2009, 71) agrees that the specific reasons for reverting to the dialectical method in the *Ph* I 2 passage and in *Cael* II 12 are different, but claims that in both cases 'the use of this procedure is again required . . . because the empirical scientific procedure is unusable'.

or cosmological tradition; rather, they arise once one is committed to the view that the heavenly bodies are carried around in concentric spheres that are themselves bodily (and not mere mathematical constructs).

Furthermore, we also believe that puzzling through the opinions of Aristotle's predecessors would be of limited value here, given that the empirical evidence in this case is limited. For, according to Aristotle, *endoxa* are valuable to the extent that they reflect or capture – perhaps in a manner that needs modification or clarification – a certain kind of relevant experience,[14] such as, in this case, *empirical* experience of cosmological phenomena.[15] This experience, of course, is mostly unavailable.[16] And even if Aristotle engages with *endoxa* in *Cael* II 12, that need not conflict with his aspirations to provide natural scientific explanations: certainly in the natural treatises, Aristotle sometimes examines *endoxa* and observational evidence *together* in order to generate scientific explanations,[17] or cites them *both* as evidence for the wrongness or rightness of certain views.[18]

Finally, the explanations that Aristotle offers in *Cael* II 12 are developed not by a priori reasoning but build on, and are held accountable to, empirical evidence from other – possibly related – domains. For, in one case, Aristotle develops an explanation with the help of an analogy with observable sublunary action: according to Aristotle, what can be observed to be the case regarding the teleological action of sublunary living beings is supposed to help us understand what is happening with the motions of the heavenly bodies, which may or may not be celestial living beings themselves. In the other case, Aristotle appeals to what is likely his own view of the teleological workings of nature as an equalizer and producer of order, for which he believes there is overwhelming empirical evidence in the sublunary realm. Thus, whatever the exact nature of these explanations may be, it is not clear to us what exactly would make them *dialectic* in nature, let alone what would make them *exclusively* dialectic.

The main evidence that Aristotle has moved away from investigating things naturally, that is, scientifically, and is investigating them dialectically, is, according to Bolton, Aristotle's appeals to *eulogos* and its cognates. Bolton considers these terms as 'one very typical and common' way in which Aristotle signals dialectical argumentation in the text. Although we agree with Bolton that Aristotle sometimes contrasts what is *eulogos* with what is apparent by observation or uses terms such as *logikôs* and *kata ton logon* for a dialectical method that is to be contrasted with a natural scientific

[14] See e.g. *NE* I 3, 1095a2–6 and *Div. Somn.* 1, 462b12–17.

[15] Cf. *APr* I 30, 46a17–22. [16] See Burnyeat (2004, 15).

[17] See e.g. *GA* I 17–18, esp. at 721b28–34. [18] See e.g. *Cael* III 4, 303a20–24.

method, we do not believe that appeals to *eulogos* and cognates are typical indicators of the dialectical method, nor that dialectic – at least in its broad sense – cannot be itself part of an empirical, scientific investigation that aims at scientific knowledge. So let us next offer a brief conceptual analysis of Aristotle's uses of the term *eulogos* and its cognates – and specifically of their uses in the natural treatises – in defence of our thesis.

Aristotle's use of the term eulogos *and its cognates in the natural treatises*

Aristotle uses the term *eulogos* and its cognates about three hundred times throughout the corpus,[19] and mostly in order to indicate that a certain thesis or phenomenon is *what can reasonably be expected*, given certain pre-established background assumptions that are spelled out in the same context, and where those background assumptions can be either dialectical or empirical in nature. In the natural treatises, these background assumptions often include facts about natural beings (e.g., about the nature of elements) or general 'laws' of nature (especially the teleology of nature, but also observed correlations as, for instance, that between body size and length of life) that Aristotle has established elsewhere and that are now invoked with regard to claims or phenomena that are potentially surprising or otherwise raise problems. And, as already noted by Bolton, Aristotle especially appeals to what is *eulogos* in empirically underdetermined domains. This is why, in the natural treatises, this type of reasoning is especially frequent in discussions of, for instance, the attributes and the locations of the various parts and functions of the soul, the powers and nature of blood and its related fluids, and the process of embryogenesis.

Consider the following example:

> Many animals do not have voice, such as for example the bloodless [animals] and among the blooded animals the fish – and this is what can reasonably be expected (καὶ τοῦτ' εὐλόγως), if indeed voice is a kind of motion of air. But those [fish] that are said to have voice, such as for example the ones in the Achelous, produce sounds with their gills or with some other such [part], but voice is the sound produced by an animal and not by any chance part. But since everything that produces voice does so by something touching something and in something, and this [something] is air, it is what can reasonably be expected (εὐλόγως) that only as many as take in air would be capable of producing voice. For nature then uses the air already breathed in for two functions; just as it uses the tongue for both tasting and articulation,

[19] Its counterpart *alogos* and its cognates occur about one hundred and fifty times. On *eulogos* in Aristotle, see also LeBlond (1938).

> and of these tasting is necessary (and so is found in a greater number (of living beings)), while expression is for the sake of well-being, so also nature uses breath both for the inner warmth, as something necessary (the reason will be stated elsewhere) and also for voice so that there may be well-being.[20]

In this passage, Aristotle addresses the potentially surprising phenomenon that some animals, such as fish, do not possess voice, even though some people claim that they do. That some animals should be voiceless, however, is exactly what is to be expected, if – as Aristotle points out – his theory about what causes voice is true, that is, if voice really is the product of air rubbing against air inside an animal, for in that case it is immediately clear that – and *why* – animals that do not take in air cannot produce voice. Aristotle does not directly appeal to empirical evidence here, as he needs to explain the *absence* of a capacity, and there is no feature that is observably present in the voiceless animals that would explain their lack of voice. But surely the explanation is a natural scientific one, as it provides a characteristically Aristotelian material causal explanation for the lack of voice in animals such as fish (i.e., since nature can only produce voice by making use of the already existing in- and outflow of air in those animals that breathe air for the sake of cooling, animals that do not breathe air lack the prerequisite material precondition for voice and thus lack voice).[21]

Where possible, however, Aristotle explicitly conjoins the conclusions one reaches when thinking about what can reasonably be expected, given how nature works always or for the most part, and what can indeed be observed. In these cases, the appeals to what is *eulogos* and what is observable are not competing strategies but complementary ways of giving the best natural scientific theories or explanations possible, given that the empirical evidence itself is underdetermined. For instance, in *PA* III 4, 667a30–b10, Aristotle claims that of all the viscera the heart alone does not undergo any severe affection (at least not without killing the animal). This claim cannot be tested – or at least not directly, without conducting vivisection – by appeal to empirical evidence, as in living beings the heart is hidden from sight. Instead, Aristotle supports it by explaining that it is *both* what is *eulogos*, given the nature of the heart as an origin (once destroyed the other parts would not be able to come to the rescue), which is something he established in the preceding section, *and* conforms to what can be observed (ὤφθαι) in sacrificial animals (healthy animals do not show

[20] *DA* II 8, 420b9–22.

[21] The *presence* of voice is thus due to *secondary teleology*: see Leunissen (2010, 68–69). Cf. *GA* IV 8, 776a25–26 for a similar example.

any signs of affections of the heart when sacrificed) and what is apparent (φαίνεται) from the dissections of sick and dying animals (these animals do exhibit morbid affections around the heart). There is no switch from dialectic method to science here, but one continuous, cumulative attempt to provide support for a biological thesis. Note, though, that because in cases like these there is no direct empirical support for a theory, it is always possible that something that now seems *eulogos* or the best account available can later – with the discovery of new empirical evidence – turn out to be false.[22] The method does not yield any absolute certainty, and even when the accounts and phenomena testify for each other, as Aristotle explains in *Cael* I 3, 270a12–b20, sometimes what can be derived from perception is only sufficient to yield a kind of 'human conviction' (ἀνθρωπίνην πίστιν).

Aristotle also uses the opposite strategy, whereby a certain theory is established as absurd (*atopos*) on the grounds that it is *both* against what can reasonably be expected (*ouk eulogos, alogos*) *and* contrary to what can be observed. This strategy is explicit in the rejection of Plato's theory of elemental transformation (*Cael* III 7, 306a1–5; 3–4: οὔτε γὰρ εὔλογον . . . οὔτε φαίνεται κατὰ τὴν αἴσθησιν). It also appears in more subtle versions elsewhere: for instance in the rejection of certain theories of reproduction of cuttlefish and crayfish (*GA* III 8, 757b31–758a21), and in the dismissal of the theory that the heavens alternate between moving slower and faster (*Cael* II 6, 289a5–8). In both cases, the difficulty of actually being able to see the relevant phenomena (because of their minuteness or our physical distance from them) means that the fact that something has not been observed has relatively little value in rejecting the theories of his competitors, as a sceptic could reply that the supporting facts for his theory have simply not been observed *yet*. Hence, by appealing to what can or cannot reasonably be expected, Aristotle is able to establish that the alternative theories are *most likely* false, given everything else we know about nature and the workings of our sense-organs.

Without trying to be exhaustive, let us briefly mention another use of *eulogos* that is common in natural science. For sometimes Aristotle claims that a potentially puzzling thesis or phenomenon is in fact what one would reasonably expect to be the case on account of it being *similar or analogous* to something else for which a causal explanation (based on empirical evidence) is readily available. Aristotle explains, for instance, that

[22] This is, for instance, how Empedocles' explanation of sexual differentation is proven to be false in *GA* IV 1, 764a33–b3.

it is *eulogos* that eggs are more prone to be spoiled in hot seasons with the following analogy:

> For just as (ὥσπερ γάρ) wines in hot conditions turn sour from the sediment getting stirred up (for this is the cause – τοῦτο γὰρ αἴτιον – of the spoiling) so it is with the yolk in eggs: for in both cases it [i.e., the sediment and the yolk] is earthy, and this is both why the wine is made turbid by having the sediment mixed in and why the eggs are spoiled [by having] the yolk [mixed in].[23]

It is clear that Aristotle does not merely offer an analogy in order to reduce the puzzlement about the phenomenon at stake; instead, he suggests that there is a similar – or perhaps even identical – causal process that is responsible for both phenomena, such that providing this analogy also yields an at least plausible causal explanation for the spoiling of eggs.[24] Note that Aristotle also uses this strategy in order to reject competing or alternative theories: a theory is *alogos* when there is *also* no empirical evidence for the parallel case[25] or when it otherwise does not fit with what is known to happen always or for the most part in nature.[26]

In other contexts, *eulogos* can be used differently, as the grounds for calling something *eulogos* can be entirely a priori or *endoxical*. In all the natural scientific cases discussed above, however, Aristotle relies on the consistency in the workings of nature, such that even limited empirical evidence can provide good reasons to accept certain explanations as plausible – as carrying at least a kind of human conviction, without constituting scientific explanations in an unqualified sense. Aristotle assumes that, if the phenomenon is to be explained at all, it must at a minimum fit with what we already know about what nature does always or for the most part; if it does not fit with this knowledge, we must defer offering an explanation of it. This is, for instance, what happens regarding the gender of bees and drones.[27] Observations of bee and drone phenotypes and behaviour do not settle which ones are male or female, and none of the available theories fit with what is otherwise known about what is typical for male and female animals. It is certainly possible that these animals form an exception, but without *strong* evidence in favour of this, Aristotle is reluctant to settle the issue. In short, something is *eulogos* if it can be shown to be epistemologically *and* ontologically consistent with, and supported by, what is

[23] *GA* III 2, 753a23–27. Cf. *GA* I 18, 725a24–27 and *GA* I 20, 729a1–14.

[24] Cf. *Cael* II 2, 284b18–24 where Aristotle explicitly employs a natural scientific principle taken from his biology to make a cosmological point via an appeal to *eulogos*.

[25] See e.g. *GA* III 11, 762b28–763a7. [26] See e.g. *GA* III 11, 759b1–7. [27] *GA* III 10, 759b1–7.

independently known about the workings of nature and helps to make the best sense of the case at hand. By reasoning about what is *eulogos*, Aristotle attempts to track (but cannot establish beyond doubt) natural scientific explanations that are true 'always or for the most part', on the general assumption that nature itself acts reasonably and hence for the best.[28]

Finally, in our review of the uses of *eulogos* and cognates in Aristotle's natural treatises, we did not find any close or typical connection with dialectic or *logikôs* methods of investigation. Instead, we found that in passages where Aristotle differentiates between investigating things *logikôs* versus investigating them *physikôs*, appeals to *eulogos* are associated with the latter and not with the former. Take, for instance, the following methodological statement:

> One could also see from these things how much those who investigate naturally (φυσικῶς) differ from those who investigate dialectically (λογικῶς):[29] for about the existence of atomic magnitudes, some say that otherwise the triangle itself would be many, whereas Democritus would appear to be convinced by *logoi* that are more appropriate and natural (οἰκείοις καὶ φυσικοῖς λόγοις). This will be clear from what we say as we go on.[30]

The passage occurs in the context of a discussion about the nature of generation and follows after a critique of people like Plato who – because of their lack of observational experience – postulate the wrong principles and make assertions too easily.[31] And while both the suggestions made by Plato and Democritus about change involving indivisible magnitudes result in consequences that are unreasonable (*aloga*), at least Democritus' suggestion is more appropriate to a natural scientific investigation than Plato's, as Aristotle's following discussion makes clear.

Now, it would be odd for Aristotle first to criticize Plato for using dialectic when it comes to physical questions, then to praise Democritus for relying on more natural scientific arguments,[32] and finally to clarify Democritus' superiority in method by relying (exclusively) on dialectical arguments himself, as allegedly signalled by his frequent appeals to what is *eulogos* or *alogos*. Instead, we believe that, as Sedley has suggested,[33] Aristotle works through Democritus' position for a second time by spelling out exactly what is (im)possible or (un)reasonable in it, but also *by reformulating*

[28] See especially *Cael* II 11, 291b11–17, but also *Cael* II 8, 290a29–31; *GA* I 23, 731a24; *GA* V 2, 781b22–23; *PA* III 2, 663a31–33; *PA* III 8, 670b32–671a1; and, perhaps, *NE* I 9, 1099b20–25.

[29] It is not certain that *logikôs* here is identical to *dialektikôs*, as Bolton believes; for a full study of the uses of *logikôs* and *dialektikôs*, see Chiba (2010, 203–51).

[30] *GC* I 2, 316a10–14. [31] *GC* I 2, 316a5–10. [32] For similar praise, see also *GC* I 8, 324b35–325a2.

[33] Sedley (2004, 67–68; 73–77).

Democritus' view in his own Aristotelian terms, and especially in terms of his distinction between potentiality and actuality. The latter distinction is a principle of his natural science;[34] it is this principle that allows Aristotle to formulate his own solution (i.e., that bodies are indivisible only in actuality, not in potentiality). The fact that Aristotle does not solve this problem directly by reference to empirical evidence (and note that it is unclear what empirical evidence could settle this particular debate) does not mean that he has given up on providing 'arguments that are more appropriate to natural science' and therefore proceeds entirely from a priori reasonings,[35] either here or in the remainder of the treatise.[36] And, finally, even though Aristotle contrasts the dialectical and the natural scientific method sharply in this passage, that does not mean that dialectical arguments (in the broad sense of the term) cannot be valuable if used *together* with empirical evidence in natural scientific investigations.[37]

Standards of explanation in empirically underdetermined domains

Aristotle is explicit about the fact that explanations and principles generated in natural investigations have to be empirical in two senses: they have to be derived at least in part from perceptible phenomena *and* at least in principle have to be answerable to them. If they do not harmonize with what is observed, one should reject them as having been wrongly assumed.[38] In empirically underdetermined domains, the explanations and principles cannot be held accountable in the same way, and so Aristotle formulates what we take to be an alternative standard for what counts as a satisfactory demonstration within natural inquiry:

> About things that are not apparent to perception, we think to have offered a satisfactory demonstration that is in accordance with reason, when we have

34 It is explicitly introduced as a first principle in, e.g., *GC* I 9, 326b29–31.

35 See e.g. *GC* I 2, 316a34–b2, 316b9–11, and 316b28–29 for appeals to empirical evidence.

36 *Pace* Burnyeat (2004, 16). Aristotle appeals to empirical evidence when possible (see e.g. *GC* II 4, 331b24–25; II 9, 335b17–24; and II 10, 336b16–18) and uses a teleological principle of nature in explaining why something is *eulogos* (*GC* II 10, 336b25–337a1).

37 See, e.g., *de Juv.* 2, 468a22–23 and *PA* III 4, 666a11–24. A passage where Aristotle rejects the *logikôs* method as not being appropriate for *any* kind of investigation is *GA* II 8, 747b27–748a16. Here the rejected method is one that searches for explanations at a level that is *too* abstract because it is *too* far removed from the objects of study and from the appropriate principles, such that it leads to false and empty explanations and fails to capture the phenomenon to be explained, but not one that is dialectical or abstract per se. For there are contexts in which something needs to be explained at exactly such an abstract level (see, e.g., *GA* III 10, 759a24–27) and certainly *logikôs* arguments have their use in natural inquiry (see, e.g. *Ph* III 5, 204b1–11).

38 See especially *Cael* III 7, 306a5–17. See also *GA* III 5, 756a2–5 and *GA* III 10, 760b28–33; cf. *GA* V 8, 788b20–24 and *Resp* 10, 476a13 on teleological principles.

brought back things to what is possible (νομίζομεν ἱκανῶς ἀποδεδεῖχθαι κατὰ τὸν λόγον, ἐὰν εἰς τὸ δυνατὸν ἀναγάγωμεν). And from the phenomena currently available one could grasp the following especially to be the case about them.[39]

Thus, when dealing with empirically underdetermined problems, Aristotle considers an explanation to be satisfactory if it cannot only be exhibited to be epistemologically possible, but also ontologically possible:[40] that is, when it is minimally consistent with and supported by the already existing body of knowledge about how nature works 'always or for the most part' in that domain, but is also the best explanation a natural scientist reasonably can provide.

The fact that there is not sufficient empirical evidence to provide a more necessary explanation for something – that is, in terms of what is either conditionally or unqualifiedly necessary – does not deter Aristotle from continuing to state what appears to be the case. The best one can do as mere mortals in such situations, Aristotle specifies, is 'to formulate suppositions that are in accordance with what can reasonably be expected (εὔλογον ὑπολαβεῖν), for saying what is necessary must be left to more powerful beings'.[41] Appeals to what is *eulogos* are thus part and parcel of Aristotle's natural scientific method that aims at the explanation of phenomena that are empirically underdetermined. However, it must still be possible to hold these appeals accountable to empirical evidence:

To claim that it [i.e., the heavens as a whole] has come to be but is nevertheless everlasting is among the impossibilities. For one should only posit those things in accordance with what can reasonably be expected that *can be observed* to be the case with regard to most or all cases (μόνα γὰρ ταῦτα θετέον εὐλόγως ὅσα ἐπὶ πολλῶν ἢ πάντων ὁρῶμεν ὑπάρχοντα), and it turns out to be the opposite concerning it [i.e., the heavens]: for all things that come to be are seen (φαίνεται) also to perish.[42]

Aristotle's argument here is that the posits we offer regarding the nature of the heavens should – as a methodological requirement – be at a minimum in agreement with what can always or for the most part be *observed* in parallel cases: even though we have no observational evidence for or against the creation or the eternality of the heavens as a whole, observations of *every other case* involving generation suggest that things that come to be

39 *Meteor* I 7, 344a5–8.

40 We thus think that the language of possibility is stronger than how Bolton (2009, 67) interprets it, who believes that it entails epistemological possibility (in addition to logical possibility) only.

41 *Meta* XII 8, 1074a14–17.

42 *Cael* I 10, 279b17–21.

also perish. Hence, the claim that the heavens have come to be but are nonetheless eternal is not a case of 'bringing back the phenomena to what is possible' and is therefore not *eulogos* and hence likely false.

Eulogos in *Cael* II 12: 'two *aporiai* about which anyone would likely be puzzled'

Introduction of the aporiai

Let us now turn to *Cael* II 12 – a methodologically rich chapter deemed to be dialectic on account of its frequent appeals to *eulogos* and cognates. This chapter contains a discussion of two *aporiai* pertaining to the motions of the heavenly bodies. Aristotle introduces the discussion as follows:

> Since there are two *aporiai* about which anyone would likely (εἰκότως) be puzzled, we have to try to say what appears to be the case (τὸ φαινόμενον), taking eagerness [to do so] to be a mark of modesty rather than boldness, if, because of a thirst for the highest form of knowledge, one is happy with even small solutions (μικρὰς εὐπορίας) concerning matters in which we have the greatest *aporiai*.[43]

Aristotle is very clear that the two *aporiai* he is about to discuss do not arise from the cosmological views defended by his predecessors or from his own particular physical theory about the motions of the heavenly bodies. Instead, Aristotle says that anyone would likely be puzzled by our *aporiai*, which presumably means: (i) *anyone* studying the heavens from a natural scientific perspective, (ii) taking the concentric spheres to be physical bodies forming a causal system, and (iii) thinking of this system in terms of order. For only such a person would likely be puzzled by the fact that the different heavenly bodies require different numbers of concentric spheres in order to move in the way they do, and specifically that there is no obvious correlation between the number of spheres and the distance to the outer sphere carrying the fixed stars (*aporia* 1), and that there is an apparent lack of order in the sense that the outer sphere carries a multitude of stars, whereas the planets are each carried separately by several spheres (*aporia* 2).

Aristotle realizes that solving these *aporiai* is not going to be an easy task, but believes that if one takes seriously the idea that even small solutions pertaining to the cosmological domain – which is, after all, honourable and divine[44] – would contribute to the highest form of knowledge, then trying

[43] 291b24–28. [44] See *PA* I 5, 644b24–25 and 645a4.

to solve these *aporiai* is simply what an eager and conscientious natural philosopher should do.[45]

So it is not really surprising that Aristotle engages in a discussion of these *aporiai*. What is remarkable is that he is very upfront about the result we should expect from his discussion. For even though they are *aporiai* for any natural philosopher searching for explanations of why things are the way they are, given that we are faced with the greatest difficulties, Aristotle says that we should be content if we can find *small solutions* (μικρὰς εὐπορίας). Accordingly, he will try to state *what appears to be the case* (τὸ φαινόμενον), and thereby perhaps will reduce our puzzlement about the different complexities of the motions of the heavenly bodies, but he will not be able to solve the *aporiai* and to offer full-fledged natural scientific explanations.

So what is then the source of the 'greatest difficulties' that keep Aristotle from solving these *aporiai*? Scholars have offered three possible answers to this question:

(1) The source of the problem is Aristotle's own lack of competence as an astronomer. His solutions to these *aporiai* are small and tentative because he lacks the technical knowledge required to offer a proper solution and also lacks the interest to properly engage with the available astronomical and mathematical knowledge (as he is merely interested in establishing the teleology of the heavens); a true astronomer would be able to give a proper solution to the *aporiai*.[46]

(2) The source of the problem is not to be found in Aristotle (except, of course, in an accidental way). *As a natural philosopher*, he does not possess the perceptual data required to give a proper solution to the *aporiai*, as he is at too great a distance from the heavens and does not have access to all the empirical information needed. In other words, physical distance entails *perceptual remoteness*, and this prevents Aristotle from solving the problem.[47]

(3) For Aristotle, the heavens are not only perceptually remote but also *conceptually* remote. The *conceptual remoteness* in question ultimately depends on the fact that the explanatory resources at our disposal are not entirely adequate to provide a positive characterization of certain important aspects of the celestial world, and this prevents Aristotle from solving the *aporiai*.[48]

[45] Cf. Aristotle's remarks in *DA* I 1, 402a1–7 and 402a10–11: 'knowing is among the fine and prized things', suggesting that one should also investigate the soul, even if 'in every way and altogether it is most difficult to gain any conviction (τινα πίστιν) concerning it'.

[46] Lloyd (1996, 160–83). [47] Leunissen (2010, 152–75). [48] Falcon (2005, 85–112).

We do not think that the first answer is correct: in *Cael*, Aristotle is not speaking as a mathematician but rather as a natural philosopher, and the difficulty of studying the heavens is a *topos* in the Aristotelian corpus.[49] Instead, we believe that a combination of the second and third answers must be the case: for *perceptual remoteness*, especially combined with *conceptual remoteness*, entails that discussions of *aporiai* pertaining to the heavenly domain will typically result in small solutions only, since many if not all of the natural scientific explanations that can be generated with the conceptual resources available to Aristotle will remain empirically underdetermined, perhaps until some further progress can be made. And, as we saw above, in such circumstances offering small solutions is entirely satisfactory for a natural philosopher, whereas not even trying to offer a solution would indicate a – for Aristotle, untypical – lack of desire for the highest form of knowledge about matters that are most divine.

This, however, presents us with a second question: to what extent do these small solutions contribute to the knowledge Aristotle calls natural philosophy? According to Bolton, the small solutions do not amount to natural scientific knowledge at all: Aristotle's modesty in this passage, combined with his appeals to what is *eulogos* in solving the *aporiai*, suggests that he has reached the limits of what can be established scientifically and that the investigation instead proceeds dialectically. Hence, the solutions do not amount to scientific explanations but are dialectical in nature.[50] We have already argued above that the mere fact that Aristotle appeals to what is *eulogos* is not necessarily an indication of dialectic. In the subsections below, we will argue that, in *Cael* II 12, Aristotle's appeal to what is *eulogos* is part of his natural scientific method, and that the small solutions Aristotle offers to the *aporiai* are scientific solutions, even though they are solutions of a weaker, less necessary kind. So let us turn to Aristotle's solutions to these *aporiai*.

The first aporia *and its 'small solution'*

Aristotle describes the first *aporia* as follows:

> Not the least amazing of the many *aporiai* that have such a character is [this]: what in the world is the cause for why it is not always the case that the [bodies] that are at a greater distance from the first motion are moved with a greater number of motions, but instead that the [bodies] that are in between [are moved with] the greatest [number of motions]? For it seems that one

[49] See especially *PA* I 5, 644b23–645a4. [50] Bolton (2009, 51–82).

> could reasonably expect (εὔλογον) that – since the first body is moved with one motion – the body nearest to it be moved with the smallest number of motions, say two, and the body next to it with three – or that there be some other such order (τάξιν). Now, the opposite happens: for the sun and the moon are moved with fewer motions than some of the wandering stars are: yet, they [i.e., the wandering stars] are further away from the centre [of the universe] and nearer to the first body than them [i.e., the sun and the moon]. With some of them, this has also become clear by observation . . . [51]

The source of the *aporia* is clear: for while one would expect to find order (τάξιν) in the celestial region pertaining to the 'distribution' of numbers of spheres to the different planets, this order is not borne out by what actually happens, or at least not in a numerical way as one might expect. That is, given that the outer stars are moved with one motion (and this, Aristotle thinks, is observable[52]), it would seem to be reasonable that the number of spheres increases proportionally relative to the distance of each planet from the outer sphere with the fixed stars. The reasonableness of the *first* expectation, that is, of there being order in the celestial region, is unrelated to anything Aristotle's predecessors might have said on the topic or with a mathematical approach to the heavens. Rather, it presupposes a natural scientific outlook that views the natural world as a complex, organized, and causally interrelated system. And this expectation is especially reasonable, of course, when one – as Aristotle does – assumes this system to be the result of the goal-directed actions of formal natures. As for the *second* expectation of *numerical* order, this is reasonable only given a mathematical approach to the study of the heavens, and this is an approach Aristotle would like to resist.[53]

In sum, Aristotle claims that it is reasonable to expect that there is *some* order in the celestial region, but that any reasons we might have for expecting *numerical* order are frustrated by what is observed to happen:[54] the sun and the moon are moved with fewer motions than *some* of the planets that are further away from us. The result is puzzlement or wonderment, of the kind that gives rise to the highest difficulties, and rightly so. For one should keep in mind that, from a natural teleological perspective, the celestial region – as a subdomain of the natural world – is expected to

[51] 291b28–292a4. [52] See *Cael* I 5, 272a5–6.

[53] Aristotle considers contemporary astronomy as a mathematical science (see e.g. *Cael* II 10, 291a29–b9; II 14, 297a2–4 and 298a15) and as not dealing with perceptible magnitudes or with the heavens above (see e.g. *APo* I 13, 78b39; *Meta* I 8, 989b33–990a15; *Meta* III 2, 997b16–998a1; and *Meta* XIII 2, 1076b39–1077a4).

[54] 292a3–9.

exhibit the highest level of order and regularity.[55] Hence, if it turns out that there is no order and regularity in the heavenly region, that would undermine Aristotle's hypothesis of order and regularity in the sublunary realm, which would ultimately undermine his theory of natural teleology. Therefore, any solution of the *aporia* must include a suggestion of how we can avoid the conclusion that there is no order in the celestial region.

Aristotle tells us that what is now puzzling should no longer appear to be 'against what can reasonably be expected' (292a17–18: οὐδὲν ἄλογον ἂν δόξειεν εἶναι), provided that we start from the following consideration:

> We tend to think of them [i.e., the heavenly bodies] merely as bodies and as units that have order but that are wholly without soul; but we ought to [think of them] under the assumption that they partake in action and life, for in that way what happens will seem to be nothing irregular (οὕτω γὰρ οὐθὲν δόξει παράλογον εἶναι τὸ συμβαῖνον).[56]

The text is ambiguous between the suggestion that we literally ought to think of the heavenly bodies *as* living beings or whether we merely ought to think of them *as if* they were living beings, as a supposition to be employed for the sake of reducing puzzlement. Following Leggatt, we adopt the '*as if*' reading.[57] We hasten to add, however, that Aristotle's invitation to think of the heavenly bodies *as if* they partook of life does not entail that for him these bodies are not in fact alive: evidence elsewhere in the Aristotelian corpus indicates that explaining the movement of the heavens requires a soul equipped with thought and desire.[58] In any case, by thinking about the heavenly bodies *in this way*, what is actually the case – namely the existing distribution of different numbers of spheres to the various planets, which at first seemed to be irregular from a numerical perspective[59] – will no longer seem unexpected (*paralogon*).[60] For, as Aristotle is about to show,

[55] See *Ph* II 4. 196a24–b5 and especially *PA* I 1, 641b10–23:

> In addition, natural science can pertain to nothing abstract, *because nature makes everything for the sake of something*... This is why it *is more likely* that the heavens have been brought into being by such a cause – if it has come to be – and is due to such a cause, than that the mortal animals have been. *Certainly the ordered and definite are far more apparent in the heavens than around us*, while the fluctuating and random are more apparent in the mortal sphere. Yet some people say that each of the animals is and comes to be by nature, while the heavens, in which there is not the slightest appearance of chance and disorder, were constituted in that way by chance and the spontaneous.

[56] 292a18–22.

[57] Cf. Leggatt (1995, 248–49). Note, however, that the reading we adopt cannot be defended on purely grammatical grounds, but follows from other exegetical assumptions about the overall strategy adopted in *Cael* II 12.

[58] On this, see further Falcon (2005, 89–97). [59] 291b31–34.

[60] Aristotle mostly uses *paralogos* to characterize events that are the opposite of what happens always or for the most part and that are therefore 'irregular', 'unexpected', or 'unusual': see *Ph* II 5,

their actual order is not *numerical* but rather *biological-axiological*: it links the hierarchical position of a (kind of) being in the *scala naturae* to the number and complexity of motions it possesses.

In inviting us to think of the heavenly bodies *as if* they were living beings, Aristotle does not invoke any of the special principles established in what we call, for lack of a better phrase, Aristotle's biology, as he does elsewhere in *Cael.*[61] Rather his view seems to be that, as long as we accept a certain analogy from everyday life, he can reduce our puzzlement concerning the first *aporia*.

The domain from which the analogy is taken is that of *goal-directed action*, which is first worked out with regard to human action. Aristotle explains that while one body may be in good shape without exercise, another body may require some exercise to be in good shape, and that the nature and amount of exercise required to be in good shape may vary. In other words, it may be more or less difficult for the second body to reach a good shape, but it will still be able to reach it. This important point is brought to our attention when Aristotle adds a case in which, for some unspecified reason, a body is permanently barred from achieving this end: no matter how much it exercises, this third body cannot ever reach a good shape. Aristotle further complicates this teleological framework by adding a case of a person who is required to do more than one thing for the sake of the same end. This person can reach the goal, to be in good shape, only by performing *a set* of actions (running, wrestling, and exercising in the dust). Aristotle notes that the higher the number of these actions are, and the more often they have to be performed, the more difficult it becomes for the person to achieve the goal. This idea is illustrated with the help of another example that is also taken from everyday life: 'to make ten thousand Chian throws playing dice is impracticable; but to make one or two throws is easy'.[62] Accordingly, by thinking of the motions of the different planets as being like these actions of different kinds of people in reaching their good, one might already start to see why the distribution of the former is not irregular.

197a18–19; *Meteor* I 12, 347b35–36; *EE* VII 14, 1247a33; *NE* V 8, 1135b16–17; and *Pol* VII 1, 1323a17–19. Aristotle rarely employs *paralogos* to characterize arguments as 'unusual' relative to what people hold commonly or for the most part: but when he does, he combines it with the term *atopos*, thereby simultaneously characterizing the argument as absurd. See e.g. *DA* I 5, 411a7–16 and *PA* I 5, 645a10–15.

61 On the use of biological principles in Aristotle's cosmology, see Lennox (2009, 187–214) and Leunissen (2010, 152–75).

62 292a28–30.

At this point, Aristotle expands the domain of goal-directed action to include that of animals and plants: 'one ought to think that the action (πρᾶξιν) of the stars too is such as that of animals and plants'.[63] The notion of action employed in this passage is that developed in the previous section: *x* is an action, if *x* is an activity oriented towards a goal that is perceived as good by the performer of *x*.[64] Nothing prevents such an action from being arranged, along with other actions, into a more or less complex package of activities oriented towards a certain goal, as Aristotle already suggested earlier with regard to human action. In the end, it is possible to conceive of a whole life as a single package of actions oriented towards a certain goal. Furthermore, depending on how complex this package is, different hierarchical levels of life are obtained. For instance, it is quite clear that human life is to be regarded as the most complex type of life, entailing a plurality of actions done for the sake of different ends (292b3–4: 'for a human being can achieve many goods, so that he can do many things, and for the sake of other things').[65] In addition, the teleological framework introduced in the previous section gives us a way to establish that the different types of life may entail different levels of access to the ultimate goal: while human beings, in spite of the complexity of their task, have in principle full access to the ultimate goal, other living beings may be barred from reaching this.

The claim that life is said in many ways, and hence includes many different types and levels of life, is arguably one of the most significant results reached in *On the Soul*. It has enormous consequences for the study of animals and plants, and could not have been taken for granted at the time.[66] And yet, Aristotle secures a very similar result in *Cael* II 12 without any explicit appeal to his psychology, or his biology, for that matter.[67]

What remains, then, is an account of how this analogy is meant to reduce the puzzlement concerning the apparent disorder in the heavenly realm (the first aporia) regarding the number of motions by which the different heavenly bodies move. The distribution of tnumber of motions does not increase numerically the further away the heavenly body is from the outer sphere: so how do we explain this unexpected, and therefore

63 292b1–2. 64 292a22–28.

65 Aristotle does not require, *expressis verbis*, that the different things done for different ends are subordinated to one ultimate end; see also below 292b8–10: 'for either there is one thing that [plants] achieve, *as also in the case of the human beings*, or if there are many things, they all contribute to the best'.

66 See Falcon (2009, 167–81).

67 Cf., however, *PA* II 10, 655b37–656a8, where a living being's position in the *scala naturae* is also connected to its complexity and number of actions.

puzzling, fact? Aristotle's solution is to think of the heavenly bodies *as if* they partake of life and action. Like actions, the motions of these bodies can be thought of as being goal-oriented. Moreover, like actions, these motions can be thought of as being ordered in more or less complex packages for the achievement of their goal. While the sphere of fixed stars achieves that goal right away with one simple motion, the planets next to it need several motions, just like some people need to engage in 'running, wrestling, and exercise in the dust' in order to maintain their fitness. But how come the sun and the moon display an apparently less complex behaviour than the other planets? In order to understand their behaviour, Aristotle appeals to the idea that there exist different levels of access to the goal based on one's hierarchical status in the *scala naturae*. Unlike the sphere of the fixed stars and the more distant planets, and just like animals, the sun and the moon do not have full access to the goal. At most, they can get close to it, and so the performance of more actions would be futile.

As we indicated earlier, Aristotle's arguments here do not rely explicitly on natural scientific principles. However, this does not make his account merely dialectic: for given that we do have solid empirical evidence for the analogous cases of the teleological actions of humans, animals, and plants,[68] Aristotle believes that we have very good natural scientific reasons for thinking that what happens in the case of the planetary motions is not as irregular as one might have thought at first (for the distribution of their motions makes sense from a *biological-axiological perspective*), even if this does not quite explain why the motions are distributed the way they are.

Making sense of the second aporia

The second *aporia* calls for an explanation of the apparent lack of another type of order in the heavens. For while a very large number of stars share one motion, namely, the simple, uniform, circular motion of the outer sphere (the sphere of the fixed stars), each of the planets (also known as wandering stars) has its own separate motions:

[68] Aristotle does not appeal *explicitly* to empirical evidence in these analogies, but surely his remarks about how much exercise is required for people to achieve health, or how easy or difficult it is to make certain throws in a game of knucklebones, or the amount of actions humans, animals, and plants are capable of performing in search of the good are supposed to capture non-theoretical facts that are apparent to everyone; otherwise the analogies would not have any explanatory force at all.

> As for the *aporia* that many stars are arranged over the first motion, which is [only] one, whereas each of the other [stars] receives its own motions separately, one could first of all[69] reasonably (εὐλόγως) believe that this occurs because of one [consideration]. For one should think with regard to life and each of the principles that there is a great superiority of the first [principle] compared to the other [principles], and that this [superiority] turns out to be in accordance with proportion (κατὰ λόγον).[70] For the first [principle], being one, moves many of the divine bodies, whereas the other [principles], being many, each moves only one [divine body]. For whichever one of the wandering stars moves with several motions. In this way, therefore, nature equalizes and produces a certain order (τινὰ τάξιν), assigning many bodies to one motion and many motions to one body.[71]

The second *aporia*, like the first one, is prompted by an apparent lack of order in the heavens to anyone approaching them from a natural scientific perspective and considers the heavens to be a causal system of concentric, bodily spheres. And here, too, Aristotle's goal is to provide a solution, albeit incomplete (μικρὰς), that reduces our puzzlement by showing that the arrangement of the heavenly bodies does display a *certain* order (τινὰ τάξιν) and that this order is *natural*. And, as a third point of similarity, the solution builds on the assumption that we should think about the heavenly bodies as partaking of life and action. It is this assumption that allows Aristotle to rephrase the second *aporia* in terms of principles of life, and to state that (i) there is a great superiority of the first principle over the other principles, and that (ii) this superiority is in accordance with some proportion (κατὰ λόγον) and therefore is *ordered*. And, by establishing this, Aristotle is able – if not to remove – at least to mitigate our puzzlement.

So how does Aristotle establish that there is some natural proportion to how bodies and motions are distributed in the heavens? He does so by invoking the principle that nature acts as an *equalizer* in assigning many motions to one body (each of the wandering stars) and one motion to many bodies (the multitude of the fixed stars). This principle looks like a variation of the general teleological principle – familiar from Aristotle's biology but emphatically not restricted to that domain – positing that nature as a goal-directed agent always does what is best, given the possibilities, and therefore always distributes or places organic parts in a balanced manner over the animal's body, thereby preserving its symmetry.[72] In this case,

69 For the second part of the solution, see our discussion of 293a4–11 below.

70 Alternatively, κατὰ λόγον can be rendered as 'in accordance with reason'.

71 292b25–293a4.

72 On such 'principles of balanced distribution', see Leunissen (2010, 124).

however, nature equalizes the heavens by assigning one motion (the superior principle) to many divine bodies and many motions (the inferior principles) to one body. Note that the notion of order introduced to solve the second *aporia* is not a *biological-axiological* one as in the first *aporia*, but rather a teleological one of balance, where the superior principle carries many, while the weaker principle needs the support of others to carry just one; neither order is simply numerical in nature as one might perhaps have expected.

In sum, the actual arrangement of the heavenly bodies over the various numbers of motions is exactly what can reasonably be expected to be the case, given our knowledge of the application of natural teleological principles in the sublunary domain. Or, to put it differently, Aristotle succeeds in reducing our puzzlement by relying on what he has established empirically in a different domain (most notably in biology).

Let us conclude by turning to the second and final part of the solution to the second *aporia*:

> And furthermore, the other motions have a single body because of this, namely that the motions prior to the last – which carries one star – move many bodies: for the last sphere moves while being embedded in many spheres, and each sphere turns out to be a body. Therefore, the work of that [final sphere] would be a common one, for while each sphere has a motion proper to it by nature, this motion [sc. the motion of the final sphere] is, as it were, added, and the capacity of every finite body is related to something finite.[73]

The second *aporia* requires an explanation of why (a) a single sphere carries a multitude of fixed stars, and why (b) a whole system of spheres jointly causes a single planet to move. More precisely, it requires an explanation of *the combination* of (a) and (b). Our passage expands on the solution of the *aporia* offered earlier, which appealed to the equalizing actions of nature, by focusing on (b). The complications of the physical explanation envisioned by Aristotle need not concern us here. What matters is that this explanation establishes, indirectly, the balance or order to be found in the distribution of 'work' to be performed by the various heavenly spheres: that is, while the outer sphere 'carries the weight' of many heavenly bodies plus one sphere (which is also a body, as Aristotle points out), the spheres carrying the planets carry one heavenly body plus many – additional, nested – spheres, so that all are doing about the same amount of work in carrying *bodies* and hence explaining the reasonableness of the observed distribution.

[73] 293a4–11.

In conclusion, Aristotle is not only a modest investigator but also one who is sensitive to the various demands of the objects under investigation. He is a modest investigator, because he presents his solutions as 'small' – as solutions that reduce our puzzlement, but that do not remove it. They are, however, scientific and not just dialectical solutions, as they are *at least in part* derived from, and are *in principle* accountable to, empirical evidence.

Bibliography

Ackrill, J. L. 1972/73. "Aristotle's Definitions of *psuchê*," *Proceedings of the Aristotelian Society* 73: 1991–33. Reprinted in Barnes, Schofield, and Sorabji (1979), 65–75.

Albritton, R. 1985. "Freedom of Will and Freedom of Action," *Proceedings and Addresses of the American Philosophical Association* 59.2: 239–51.

Alexander of Aphrodisias. 1887. *De Anima cum Mantissa*, I. Bruns (ed.), Berlin: Reimer.

Balme, D. 1972 [1992]. *Aristotle. De Partibus Animalium I and De Generatione Animalium I (with Passages from II.1–3). Translated with Notes. With a Report on Recent Work and an Additional Bibliography*, Oxford: Clarendon Press.

1987. "Teleology and Necessity," in A. Gotthelf and J. Lennox (eds.), *Philosophical Issues in Aristotle's Biology*, Cambridge: Cambridge University Press, 275–85.

Barnes, Jonathan. 1971. "Aristotle's Concept of Mind," *Proceedings of the Aristotelian Society* 72: 101–14.

(ed.). 1984. *The Complete Works of Aristotle: The Revised Oxford Translation*, Princeton: Princeton University Press.

(tr.). 1993. *Aristotle: Posterior Analytics* (second edition), Oxford: Clarendon Press.

Barnes, J., Schofield, M., and Sorabji, R. 1975. *Articles on Aristotle, Vol. I: Science*, London: Duckworth.

Barnes, J., Schofield, M., and Sorabji, R. 1979. *Articles on Aristotle, Vol. IV: Psychology and Aesthetics*, London: Duckworth.

Bodnar, I. 2005. "Teleology across Natures," *Rhizai* 2: 9–29.

Bolton, R. 1976. "Essentialism and Semantic Theory in Aristotle: *Posterior Analytics* II, 7–10," *Philosophical Review* 85: 514–44.

1978. "Aristotle's Definitions of the Soul: *De Anima* II, 1–3," *Phronesis* 23: 258–78.

1997. "The Material Cause: Matter and Explanation in Aristotle's Natural Science," in *Aristotelische Biologie*, Stuttgart: Steiner, 97–126.

2009. "Two Standards for Inquiry in Aristotle's *De caelo*," in A. C. Bowen and Ch. Wildberg (eds.), *New Perspectives on Aristotle's* De caelo, Leiden: Brill, 51–82.

2010. "Biology and Metaphysics in Aristotle," in R. Bolton and J. G. Lennox (eds.), *Being, Nature, and Life in Aristotle*, Cambridge: Cambridge University Press, 30–55.
Bradie, M. and Miller, F. 1984. "Teleology and Natural Necessity in Aristotle," *History of Philosophy Quarterly* 1: 133–45.
Burnyeat, M. F. 2004. "Introduction: Aristotle on the Foundations of Sublunary Physics," in F. A. J. de Haas and J. Mansfeld (eds.), *Aristotle's On Generation and Corruption* I *Book I, Symposium Aristotelicum*, Oxford: Oxford University Press, 7–24.
Bywater, I. (ed.). 1894. *Aristotelis* Ethica Nicomachea. Oxford: Oxford University Press.
Caston, V. 2009. "Commentary on Charles," *Proceedings of the Boston Area Colloquium in Ancient Philosophy* 24: 30–49.
Charles, D. 1988. "Aristotle on Hypothetical Necessity and Irreducibility," *Pacific Philosophical Quarterly* 69: 1–53.
1991. "Teleological Causation in the *Physics*," in L. Judson (ed.), *Aristotle's Physics: A Collection of Essays*, Oxford: Oxford University Press, 101–28.
2000. *Aristotle on Meaning and Essence*, Oxford: Clarendon Press.
2009. "Aristotle's Psychological Theory," *Proceedings of the Boston Area Colloquium in Ancient Philosophy* 24: 1–30.
Charlton, W. 1992. *Aristotle's Physics. Books 1 & 2; Translated [from the Greek] with Introduction and Notes*, Oxford: Clarendon Press.
Chiba, K. 2010. "Aristotle on Essence and Defining-Phrase in his Dialectic," in D. Charles (ed.), *Definition in Greek Philosophy*, Oxford: Oxford University Press, 203–51.
Classen, C. J. 1981. "Aristotle's Picture of the Sophists," in G. Kerferd (ed.), *The Sophists and Their Legacy*, Wiesbaden: Franz Steiner.
Cohen, S. 1989. "Aristotle on Hot, Cold, and Teleological Explanation," *Ancient Philosophy* 9: 255–70.
Cooper, J. 1982. "Aristotle on Natural Teleology," in *Language and Logos*, Cambridge: Cambridge University Press, 197–222; reprinted in Cooper (2004), *Knowledge, Nature, and the Good: Essays in Ancient Philosophy*, Princeton: Princeton University Press, 107–29.
1987. "Hypothetical Necessity and Natural Teleology," in Gotthelf and Lennox (eds.), 243–74.
Corcilius, K. and Gregoric, P. (2010) "Separability vs. Difference: Parts and Capacities of the Soul in Aristotle," *Oxford Studies in Ancient Philosophy* 39: 81–119.
Crisp, R. (tr.). 2000. *Aristotle: Nicomachean Ethics*. Cambridge: Cambridge University Press.
Demoss, D. and Devereux, D. 1988. "Essence, Existence, and Nominal Definition in Aristotle's *Posterior Analytics* II 8–10," *Phronesis* 33: 133–54.
Denniston, J. D. 1950. *The Greek Particles* (second edition), Oxford: Clarendon Press.
Detel, W. 1993. *Aristoteles, Analytica Posteriora*, Berlin: Akademie Verlag.
Ebrey, D. 2007. "Aristotle's Motivation for Matter" (UCLA PhD Dissertation).

2014. "Making Room for Matter: Material Causes in the *Phaedo* and the *Physics*," *Apeiron* 47: 245–65.

Everson, S. 2007. "The *De Somno* and Aristotle's Explanation of Sleep," *Classical Quarterly* 57: 502–20.

Falcon, A. 2005. *Aristotle and the Science of Nature Unity without Uniformity*, Cambridge: Cambridge University Press.

2009. "Aristotle on the Scope and Unity of the *De Anima*," in G. van Riel and P. Destrée (eds.), *Ancient Perspectives on Aristotle's De anima*, Leuven: Leuven University Press, 167–81.

Fowler, H. N. (tr). 1925. *Plato: Statesman*, Cambridge, MA: Harvard University Press.

Freeland, C. 1987. "Aristotle on Bodies, Matter, and Potentiality," in Gotthelf and Lennox (eds.), 392–407.

Frey, C. 2007. "Organic Unity and the Matter of Man," *Oxford Studies in Ancient Philosophy* 32: 167–204.

2015. "From Bloal to Flesh: Homonymy, Unity, and Ways of Being in Aristotle" *Ancient Philosophy*, 35(2).

Furley, D. 1985. "The Rainfall Example in *Physics* II 8," in A. Gotthelf (ed.), *Aristotle on Nature and Living Things*, Pittsburgh: Mathesis Publications, 177–82.

Gaiser, K. 1969. "Das zweifache Telos bei Aristoteles," in I. Düring (ed.), *Naturphilosophie bei Aristoteles und Theophrast*, 4th Symposium Aristotelicum, Heidelberg: Lothat Stiehm Verlag, 97–113.

Gallop, D. 1991. *Aristotle. On Sleep and Dreams*, Warminster, UK: Aris and Phillips.

Gauthier, R. A. and J. Y. Jolif. 1970. *Aristotle*: L'Ethique a Nicomaque. 3 Vols. Louvain and Paris.

Gigon, O. 1987. *Librorum deperditorum fragmenta = vol.* III *Aristoteles Opera*, Berlin: De Gruyter.

Gill, M. L. 1997. "Material Necessity and *Meteorology* vi 12," in W. Kullmann and S. Follinger (eds.), *Aristotelische Biologie: Intentionen, Methoden, Ergebnisse*, Stuttgart: Franz Steiner Verlag, 145–62.

Gotthelf, A. 1976. "Aristotle's Conception of Final Causality," *Review of Metaphysics* 30: 226–54; reprinted with additional notes and postscript in Gotthelf (2012) *Teleology, First Principles, and Scientific Method in Aristotle's Biology*, Oxford: Oxford University Press, 3–44.

(ed.). 1985. *Aristotle on Nature and Living Things*, Pittsburgh: Mathesis Publications.

1985 [2012]. "Notes Towards a Study of Substance and Essence in Aristotle's *Parts of Animals* ii–iv," in Gotthelf (ed.), *Aristotle on Nature and Living Things*, 27–54. Reprinted in Gotthelf (2012), Ch. 10.

1987. "Aristotle's Conception of Final Causality: With Postscript," in Gotthelf and Lennox (eds.), 204–42.

1987 [2012]. "First Principles in Aristotle's *Parts of Animals*," in Gotthelf and Lennox (eds.), 204–42, reprinted in Gotthelf (2012), Ch. 7.

1997. "Understanding Aristotle's Teleology," in R. F. Hassing (ed.), *Final Causality in Nature and Human Affairs*, Washington DC: Catholic University of America Press, 71–82.

1997 [2012]. "The Elephant's Nose: Further Reflections on the Axiomatic Structure of Biological Explanation in Aristotle," in *Aristotelische Biologie*, Stuttgart: Steiner, 85–96; reprinted in Gotthelf (2012), Ch. 8.
2012. *Teleology, First Principles, and Scientific Method in Aristotle's Biology*, Oxford: Oxford University Press.
Gotthelf, A. and Lennox, J. 1987. *Philosophical Issues in Aristotle's Biology*, Cambridge: Cambridge University Press.
Gregoric, P. 2007. *Aristotle on the Common Sense*, Oxford: Oxford University Press.
Hamlyn, D. W. 2002. *Aristotle De Anima Books* II *and* III, Oxford: Clarendon Press.
Hankinson, J. 2005. "Aristotle on Kind-Crossing," in R. W. Sharples (ed.), *Philosophy and the Sciences in Antiquity*, Aldershot: Ashgate, 23–54.
Hardie, R. P. and Gaye, R. K. (tr.). 1984. *Aristotle: Physics*, in Barnes (ed.), Vol. 1, 315–446.
Heinaman, R. 1997. "Frede and Patzig on Definition in *Metaphysics* Z.10–11," *Phronesis* 42: 283–98.
Henry, D. 2007. "How Sexist is Aristotle's Developmental Biology?", *Phronesis* 52: 251–69.
2008. "Organismal Natures," *Apeiron* 3: 47–74.
2013. "Optimality Reasoning in Aristotle's Natural Teleology," *Oxford Studies in Ancient Philosophy* 45: 225–63.
Hicks, R. D. 1907. *Aristotle De Anima*, Cambridge: Cambridge University Press.
Hutchinson, D. S. and Johnson, M. R. (tr.). *Aristotle: Protrepticus* (www.protrepticus.info).
Inwood, B. and Woolf, R. (tr.). 2013. *Aristotle: Eudemian Ethics*, Cambridge: Cambridge University Press.
Irwin, T. 1988. *Aristotle's First Principles*, Oxford: Oxford University Press.
(tr.). 1999. *Aristotle: Nicomachean Ethics*. Indianapolis and Cambridge, MA: Hackett Publishing Company.
Jaeger, W. 1948 [1961]. *Aristotle: Fundamentals of the History of his Development*, R. Robinson (tr.), Oxford: Oxford University Press. Originally published in German (first edition 1923; second edition 1955), *Aristoteles: Grundlegung einer Geschichte seiner Entwicklung*, Berlin.
Joachim, H. 1926. *Aristotle On Coming-to-be and Passing-away, Greek text with commentary*, Oxford: Clarendon Press.
Johansen, T. 2012. *The Powers of Aristotle's Soul*, Oxford: Oxford University Press.
Johnson, M. R. 2005. *Aristotle on Teleology*, Oxford: Oxford University Press.
2009. "The Aristotelian Explanation of the Halo," *Apeiron* 42: 325–57.
2011. Review of Bolton and Lennox 2010, *Notre Dame Philosophical Reviews* 9/21.
Judson, L. 2005. "Aristotelian Teleology," *Oxford Studies in Ancient Philosophy* 29: 341–66.
Kahn, C. 1985. "The Place of the Prime Mover in Aristotle's Teleology," in A. Gotthelf (ed.), *Aristotle on Nature and Living Things*, Pittsburgh: Mathesis Publications, 183–205.

Kullman, W. 1985. "Different Concepts of the Final Cause in Aristotle," in A. Gotthelf (ed.), *Aristotle on Nature and Living Things*, Pittsburgh: Mathesis Publications, 169–75.

Kung, J. 1982. "Aristotle's *De Motu Animalium* and the Separability of the Sciences," *Journal of the History of Philosophy* 20: 65–76.

LeBlond, J.-M. 1938. *Eulogos et l'argument de convenance chez Aristote*, Paris: Les Belles Lettres.

Leggatt, S. 1995. *Aristotle On the Heavens I & II*, Warminster: Aris& Phillips.

Lennox, J. 1985. "Are Aristotelian Species Eternal?", in A. Gotthelf (ed.), *Aristotle on Nature and Living Things*, Pittsburgh: Mathesis Publications, 67–94.

1986. "Aristotle, Galileo, and 'Mixed Sciences'," In W. A. Wallace (ed.), *Reinterpreting Galileo*, Washington DC: Catholic University of America Press, 29–51.

1987. "Divide and Explain: the *Posterior Analytics* in Practice," in Gotthelf and Lennox (eds.), 90–119.

2001a. *Aristotle: On the Parts of Animals* I–IV, Oxford: Oxford University Press.

2001b. *Aristotle's Philosophy of Biology: Studies in the Origins of Life Science*, Cambridge: Cambridge University Press.

2001c. "Material and Formal Natures in Aristotle's *De Partibus Animalium*," in Lennox 2001a, 182–204.

2009. "*De Caelo* 2.2 and its Debt to *De Incessu anImalium*," in A. C. Bowen and Ch. Wildberg (eds.), *New Perspectives on Aristotle's De caelo*, Leiden: Brill, 187–214.

2010. "The Unity and Purpose of *On the Parts of Animals I*," in J. Lennox and R. Bolton (eds.), *Being, Nature, and Life in Aristotle*, Cambridge: Cambridge University Press, 56–77.

Lennox, J. and Bolton, R. (ed.). 2010. *Being, Nature, and Life in Aristotle*, Cambridge: Cambridge University Press.

Leszl, W. 1970. *Logic and Metaphysics in Aristotle*, Padua: Antenore.

Lettinck, P. 1999. *Aristotle's Meteorology and its Reception in the Arab World*, Leiden: Brill Academic Publishers.

Leunissen, M. 2010. *Explanation and Teleology in Aristotle's Science of Nature*, Cambridge: Cambridge University Press.

Lewis, F. 1988. "Teleology and Material/Efficient Causes in Aristotle," *Pacific Philosophical Quarterly* 69: 54–98.

Liddell, H. G., Scott, R., and Jones, H. S. 1996. *A Greek–English Lexicon* (ninth edition, revised), Oxford: Clarendon Press.

Lloyd, G. E. R. 1962. "Genus, Species, and Ordered Series in Aristotle," *Phronesis* 18, 67–90.

1996. "Heavenly Aberrations," in G. E. R. Lloyd, *Aristotelian Explorations*, Cambridge: Cambridge University Press, 160–83.

Locke, J. 1975. *An Essay concerning Human Understanding*, P. H. Nidditch (ed.), Oxford: Oxford University Press.

Lowe, M. 1978. "Aristotle's *De Somno* and His Theory of Causes," *Phronesis* 23: 279–91.

Lulofs, Drossart H. J. 1947. *Aristotelis De Insomniis et De Divinatione per Somnum*, Leiden: Brill.

Malink, M. 2013. "Essence and Being: A Discussion of Michael Peramatzis, *Priority in Aristotle's Metaphysics*," *Oxford Studies in Ancient Philosophy* 45: 341–62.

Matthews, G. 1992. "*De Anima* ii, 2–4 and the Meaning of Life," in M. Nussbaum and A. Rorty (eds.), *Essays on Aristotle's De Anima*, Oxford: Clarendon Press.

McKirahan, R. D. Jr. 1978. "Aristotle's Subordinate Sciences," *The British Journal for the History of Science* 11: 197–220.

Menn, S. 2002. "Aristotle's Definition of Soul in the *De Anima*," *Oxford Studies in Ancient Philosophy* 22, Summer: 83–140.

Morel, P.-M. 2006. "'Common to Soul and Body' in the *Parva Naturalia*," in R. A. H. King (ed.), *Common to Body And Soul: Philosophical Approaches to Explaining Living Behaviour*, Berlin: Clarendon Press, 121–39.

Newman, J. H. 1903. *An Essay in Aid of a Grammar of Assent*, London: Longmans, Green and Co.

1906. *Loss and Gain*, London: Longmans, Green and Co.

1907a. *Discussions and Arguments on Various Subjects*, London: Longmans, Green and Co.

1907b. *Parochial and Plain Sermons*, 8 vols., London: Longmans, Green and Co.

1909. *Fifteen Sermons Preached before the University of Oxford*, London: Longmans, Green and Co.

Norlin, G. (tr.) 1928. *Isocrates: Antidosis*, in *Isocrates*, Vol. II, Cambridge, MA: Harvard University Press, 179–366.

Nussbaum, M. 1978. *Aristotle's De Motu Animalium: Text with Translation, Commentary, and Interpretive Essays*, Princeton: Princeton University Press.

Owen, G. E. L. 1960. "Logic and Metaphysics in Some Earlier Works of Aristotle," in I. Düring and G. E. L. Owen (eds.), *Aristotle and Plato in the Mid Fourth Century*, Göteborg: Almqvist and Wiksell.

1970. "Aristotle: Method, Physics, Cosmology," in C. C. Gillespie (ed.), *Dictionary of Scientific Biography*, Vol. I, 250–58. Reprinted in G. E. L. Owen (1986), *Logic, Science, and Dialectic. Collected Papers in Greek Philosophy*, Ithaca, NY: Cornell University Press, 151–64.

Peck, A. L. 1937. *Aristotle: Parts of Animals*, Translation and Commentary. Loeb. Cambridge, MA: Harvard University Press.

1942. *Aristotle: Generation of Animals*, Translation and Commentary. Loeb. Cambridge, MA: Harvard University Press.

Pellegrin, P. 2005. *Aristote, Seconds Analytiques*, Paris: Flammarion.

Pickard-Cambridge, W. A. (tr.). 1984. *Aristotle: Topics*, in J. Barnes (ed.), *The Complete Works of Aristotle*, Vol. I, Princeton: Princeton University Press, 167–277.

Polansky, R. 2007. *Aristotle's De anima*, Cambridge: Cambridge University Press.

Quarantotto, D. 2005. *Causa finale, sostanza, essenza in Aristotele: Saggi sulla struttura dei processi teleologici naturali e sulla funzione del Telos*, Naples: Bibliopolis.

Rackham, H. 1926 [1934]. *Aristotle: Nicomachean Ethics* (revised edition), Cambridge, M A: Harvard University Press.

Rashed, M. 2005. *Aristote. De la géneration et la corruption. Nouvelle édition*. Paris: Les Belles Lettres.
Robinson, R. (tr.). 1962. *Aristotle: Politics Books* III *and* IV, Oxford: Clarendon Press.
Rodier, G. 1900. *Aristote Traite de l'Âme*, Paris: E. Leroux.
Rosen, J. 2008. "Necessity and Teleology in Aristotle's *Physics*" (Princeton PhD Dissertation).
Ross, W. D. 1924. *Aristotle's Metaphysics: A Revised Text with Introduction and Commentary*, Vol. II, Oxford: Clarendon Press.
(tr.). 1925. *Nicomachean Ethics*, in *The Works of Aristotle Translated into English under the Editorship of W. D. Ross*, Vol. IX. Oxford: Clarendon Press.
1949. *Aristotle, Prior and Posterior Analytics*, Oxford: Clarendon Press.
1955. *Aristotle. Parva Naturalia, a Revised Text with Introduction and Commentary*, Oxford: Clarendon Press.
(tr.). 1984a. *Aristotle: Metaphysics*, in J. Barnes (ed.), *The Complete Works of Aristotle*, Vol. II, Princeton: Princeton University Press, 1,552–728.
(tr.). 1984b. *Aristotle: Nicomachean Ethics*, in J. Barnes (ed.), *The Complete Works of Aristotle*, Vol. II, Princeton: Princeton University Press, 1,729–867. Revised by J. O. Urmson.
Rowe, C. J. (tr.). 1997. *Plato: Statesman*, in J. Cooper and D. S. Hutchinson (eds.), *Plato: Complete Works*, Indianapolis: Hackett, 294–358.
(tr.). 2002. *Aristotle: Nicomachean Ethics*, Oxford: Oxford University Press.
Saunders, T. J. (tr.). 1995. *Aristotle: Politics Books* I *and* II, Oxford: Clarendon Press.
Sauvé Meyer, S. 1992. "Aristotle, Teleology, and Reduction," *Philosophical Review* 101: 791–825.
Scharle, M. 2008. "Elemental Teleology in *Physics* II.8," *Oxford Studies in Ancient Philosophy* 34: 147–83.
Sedley, D. 1991. "Is Aristotle's Teleology Anthropocentric?", *Phronesis* 36: 179–96.
2007. *Creationism and Its Critics in Antiquity*. Sather Classical Lectures, 66, Berkeley: University of California Press.
2000. "*Metaphysics* Λ.10," in M. Frede and D. Charles (eds.), *Aristotle's Metaphysics Lambda, Symposium Aristotelicum*, Oxford: Oxford University Press, 327–50.
2004. "*On Generation and Corruption* II 2," in F. A. J. de Haas and J. Mansfeld (eds.), *Aristotle's On Generation and Corruption, Book I, Symposium Aristotelicum*, Oxford: Oxford University Press, 65–89.
2010. "Teleology: Aristotelian and Platonic," in J. G. Lennox and R. Bolton (eds.), *Being, Nature, and Life in Aristotle*, Cambridge: Cambridge University Press, 5–29.
Segev, M. 2012. "The Teleological Significance of Dreaming in Aristotle," *Oxford Studies in Ancient Philosophy* 43: 107–41.
Shields, C. 2002. *Order in Multiplicity: Homonymy in the Philosophy of Aristotle*, Oxford: Oxford University Press.
Simplicius. 1882. *In Libros Aristotelis De Anima Commentaria*, M. Hayduck (ed.), Berlin: Reimer.
Sorabji, R. 1974. "Body and Soul in Aristotle," *Philosophy* 49, 63–89.

Trendelenburg, F. A. 1877. *Aristotelis De Anima Libri Tres*, Berlin: W. Weber.

von Fritz, K. and Kapp, E. 1950. *Aristotle's Constitution of Athens and Related Texts*, New York: Hafner Press.

Ward, J. 1996. "Souls and Figures: Defining the Soul in *De Anima* II 3," *Ancient Philosophy* 16: 113–28.

Waterlow [Broadie], S. 1982. *Nature, Change, and Agency in Aristotle's* Physics, Oxford: Oxford University Press.

Whiting, J. 2002. "Locomotive Soul," *Oxford Studies in Ancient Philosophy* 22: 141–200.

Wiesner, J. (1978), "The Unity of the Treatise *De Somno* and the Physiological Explanation of Sleep in Aristotle," in G. E. R. Lloyd and G. E. L. Owen (eds.), *Aristotle on Mind and the Senses: Proceedings of the Seventh Symposium Aristotelicum*, Cambridge: Cambridge University Press, 241–80.

Wilson, M. 2014. *Structure and Method in Aristotle's* Meteorologica, Cambridge: Cambridge University Press.

Zabarella, J. 1594. *In Duos Aristotelis Libros Posteriores Analyticos Commentarii*, Basel.

Index locorum

General index